RICECRAFT

BY MARGARET GIN

ILLUSTRATED BY WIN NG

YERBA BUENA PRESS

SAN FRANCISCO

1975

To the Good Mother
Earth, whose bounty and
benevolence we give
our thanks

ISBN 0912738-05-7

Library of Congress Card #74-14290
Printed in The United States Of America
Copyright © 1975 by Margaret Gin
Published by Yerba Buena Press
666 Howard St.
San Francisco, California 94105
First Edition

Distributed By Random House, Inc.
and in Canada By Random House of Canada, Ltd.
ISBN 0-394-73050-X

About the author

Margaret Gin was born in Tucson, Arizona in 1933. It was here that her Cantonese parents settled and operated a grocery store. It was here, too, against a backdrop of American, Mexican and Cantonese cultural ingredients that Maggi learned to cook and sew.

Showing a rare talent in both these arts, she eventually came to San Francisco to design ski wear. And, being a young woman of boundless energy and enthusiasm, she soon found herself presiding over sumptuous Oriental feasts, for which she fast became famous.

Married to free-lance commercial artist, William Gin, and mother of two teen-age sons, Robert and William, Maggi enjoys cooking, collecting antiques, partaking in the musical life of San Francisco and relaxing in their country home in the beautiful Napa Valley.

At the urging of her admiring friends, Maggi taught Chinese cooking classes and began writing cook books. She co-authored *Peasant Cooking of Many Lands* with Coralie Castle and *Innards and Other Variety Meats* with Jana Allen.

Her most recent work, *Ricecraft,* is a natural. Rice has invariably played a vital role in her life since childhood — "There wasn't a day when a bowl of rice didn't appear on the table". And rice has the special characteristics that Maggi Gin appreciates and promotes: value, variety, taste and excitement.

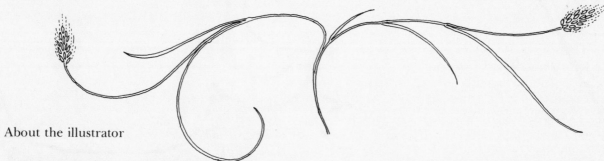

About the illustrator

Win Ng, illustrator of *Herbcraft, Worcraft, Plantcraft, Eggcraft,* and *Teacraft,* lends his special talents to yet another craft compendium. Win Ng was born and raised in San Francisco, is a master ceramicist and creative retailer. Whatever the endeavor, the results brim with his entertaining genius and enthusiasm.

CONTENTS

INTRODUCTION

Rice, one of the world's oldest and most cultivated grains, has a long and fascinating history that continues to unfold as scientists explore its secrets and cooks discover its versatility — not only as a grain that enhances any meal, but one that is fulfilling, healthy, comforting and allergy-free. It is economical and easy to prepare and store as well as being a fuel-saving food.

Rice blends with foods as it has blended with cultures. Its production has consumed the energies of nations and millions of people. Each country has its own rice specialty — the colorful *paellas* of sunny Spain, the creamy *risottos* of Italy, the pearl smoothness of the pressed rice *sushi* of Japan, the fried rice of China to the spicy *Jambalaya* of the American South. Even plain steamed rice is in its glory when presented in Dutch Rijsttafel or Indian curries.

What this food is, where it comes from, how it is used throughout the world and how to make rice a part of your daily feast are adventures that *Ricecraft* pursues with gusto and success.

OF RICE & MEN

It is in the hand of the farmer that the rice plant grows
Old Chinese Saying

Early man, like most other creatures in the animal world, foraged wide and far for food in field and forest, in stream and swamp, and in the sea. Wandering tribes took little heed of social structure or land ownership, for they sought sustenance wherever Nature was kind enough to provide it and rarely was there food enough in one spot to support very many humans.

Historians believe it was the Bengalese, about 3000 B.C., who first learned that the spindly wild grain they found in the hot moist swamps of Southeast Asia could be cultivated, thereby providing a more stable and bountiful source of food. No one knows how many years it took to transform wild grass into the lush,

productive plant we know today with its fat, pearly kernels containing valuable proteins, minerals and vitamins and rich in digestible carbohydrates. But there can be no doubt that rice revolutionized man's social and economic history.

From seeding, transplanting, fertilizing, weeding and cultivating to harvesting, rice truly does grow in the hand of the diligent farmer — not individual farmers scattered thinly over large areas, but tightly knit groups laboring shoulder to shoulder in steamy rice paddies, living close by in organized communities. No longer nomadic, man became a social creature with a need for rules and customs regarding how to live together and share the precious land that had become so productive. Government developed, and property rights became so important that wars were fought to determine ownership. Finding adequate food was no longer an all-consuming task, however. There was now time to contemplate man's nature, to create man-made beauty, to study the workings of the physical world. The arts and sciences were born.

Little wonder that rice is considered sacred by many even today — a precious common bond with God and earth — or that rice growing has engendered so many ceremonies and legends. In Japan, individual rice fields are given names as if they were persons. The Javanese say rice, because it has a soul, must never be stolen even in the direst circumstances. Legend has it that the Hindu god, Shiva, seeking the favors of the goddess Retna Dumila, found she would never submit unless he created the perfect food. His lieutenant, Kala Gumerog, was faced with a similar ultimatum from his enamored, Dewie Sri (now goddess of rice). Both failed. Virgins to their deaths, the two enchantresses were buried by their sorrowing suitors. From their tombs sprang the perfect food: rice.

The Chinese date their first cultivation of rice back 5,000 years, when the father of Chinese agriculture, Emperor Shen Nung, held yearly planting ceremonies. Only he could plant the finest seeds of this most sacred of grains. Princes and lower ranking officials were then allowed to sow soy beans, millet, wheat and barley, followed by the peasants and their planting. Even today rice planting in the Orient is a time of jubilant celebrations. Every June 14 in Osaka, for example, a dozen country girls are chosen to perform the ceremony of transplanting seedlings in the paddy fields of Sumihoshi Shrine. The Japanese word for rice, *gohan*, also is used for "meal", for no meal can be complete without rice. "How do you do?" becomes "Have you eaten your rice?" as a greeting. In China "breaking bread" is replaced by "eating rice".

Although rice thrown at brides was originally a prayer for fertility, one story tells of a bride pursued by a huge golden pheasant sent by a sorceror to destroy her. Rice was thrown out the door, and as the great bird pecked away the bride escaped in her red bridal chair. In Madagascar, throwing rice or playing a flute while rice still stood in the fields was sure to bring a destructive hailstorm. Rice grains placed on an Arab's roof preserved all from misfortune. An ancient Chinese remedy for aching bones, runny noses and stomach aches called for toasted brown rice and chopped ginger root stir-fried in whiskey, then tied in a handkerchief to rub on aching bones or stomach, or to inhale the hot ginger essence. Uncooked rice has had its uses over the ages, too, as an abrasive cleanser. A handful of rice swished inside a violin collects dust from the crevices; a few grains shaken in baby bottles made them shine.

By the Fourth Century B.C. rice had reached Egypt by way of Persia, where *chilau* is still considered the national dish of modern Iran. Meat, eggs, vegetables and fruits are considered only satellite foods as they are in so many other countries where rice is king. The Moors introduced rice to Spain, whose colonizers carried it across the sea to Central and South America as it spread by land to other parts of Europe. The justly famous *Risotto Milanese* was born in Italy's Po Valley, where rice growing began in the 15th Century. By the end of the 18th Century, Italy's production was of such national importance it was illegal to take seeds out of the country, though it is reported Thomas Jefferson smuggled some out in his pockets for planting at Monticello.

In 1686 a ship from Madagascar landing in Charleston, South Carolina, for repairs was received with typical southern hospitality. To show his appreciation the captain presented a bag of rice to one

of the city fathers who promptly planted the seeds in a nearby swamp with a yield almost sufficient to feed the entire colony. America was now in the rice business, and the British, who had previously shown little interest in Mediterranean varieties, began to import rice by the shipload. Louisiana, Texas, Arkansas and California have since become the major American rice producers (two billion pounds each in 1973), and American agriculturalists have recently developed a new "super" strain to help feed the hungry millions in Asia for whom rice is the staff of life.

Thus rice has circled the earth through the centuries. No grain yields more sustenance per acre of land, yet no crop requires more labor for its cultivation. More workers are needed to produce more rice; more rice permits population growth, which provides more workers. Like a dog chasing its tail, man seems to go around in circles in his quest for a balance between population and food. There are both dire and optimistic predictions of the outcome. Only one thing is certain: just as rice has played a vital role in man's past, it will continue to do so in his future.

CULTURE OF THE KERNEL

Over 7,000 varieties of *Oryza sativa*, common rice, usually in the form of boiled whole kernels, feed approximately as many of the world's people as do all kinds of wheat ground into flour and baked into bread. Although well under one-half as many acres of land need be cultivated to produce such prodigious quantities of rice as compared to wheat, it is estimated that more than 100 times as many man-hours are required, for most of the world's crop is still planted, cultivated and harvested by hand just as it was many centuries ago.

Though boiling rice is far simpler than grinding flour and baking bread, the growing process is much more complex than that of wheat. Except for the relatively insignificant upland varieties that require dry ground, rice demands flooded level fields or terraces, taking all of its oxygen from its leaves through specialized air-conducting tissues. Plenty of sunlight and warm temperatures (60° to 80°) are needed; soil fertility is not always critical.

The rice plant grows from two to five feet tall, depending on the variety and depth of submersion, and is characterized by perfect flowers bearing six stamens and a single pistil. Grain is produced on nodding panicles of spikelets and when ripe resembles oats. Like other cereals, each grain is composed of a small germ enclosed in an endosperm of mostly starch and certain vitamins and minerals. Several brown, tight layers of bran outer coatings serve as protection for the embryo and its initial food supply, and the kernel is nestled within a loose, multi-leaved hull.

Sowing thickly in seedbeds, followed by transplanting of seedlings in paddies of mud, reduces the growing period to a minimum — in some tropical areas as short as 90 days. This method is so tedious that other practices have been developed, especially in the United States, such as using mechanized seed drills on fields before flooding or even broadcasting seed over large fields by airplane. Once the plants are established, fields may be drained for weeding, cultivating and fertilizing and subsequent reflooding. This process may be repeated as many as six times during a single growing season to keep the fields free of foreign material and to stimulate growth.

At harvest time, after fields are drained and allowed to dry, workers or machines can reap the crop like other grains. In the Orient, hand sickles are usually used to cut the stalks, which are then bound by hand into sheaves and stacked for drying. In Java, a small bird-shaped knife called "ani ani" makes cutting rice stalks even more laborious, but since this knife is sacred to the rice goddess, Dewi Sri, more efficient tools are shunned.

Reaping and threshing call for jubilant celebrations, for the precious food whole families have labored together to produce is at last available to fill all the hungry mouths. In Thailand, as the grain is ripening the farmers place bamboo poles in paddies, not only to signify that Mother Posop, their rice goddess, is pregnant but also to ward off evil spirits. The new crop is blessed and spoken to with great love and admiration during harvesting and threshing. Some of the new

grain is ceremoniously mixed with the special seed rice to ensure growth of the next planting.

Golden sheaves are ready to thresh when not so dry that kernels will crack, but dry enough for the kernels to be readily knocked off the stalks by beating over a barrel or drawing across steel-toothed tools. Further sun drying on large mats is required before hulls can be pounded loose from the kernels. Winnowing, or removing the chaff from the grain, is said to be best done by woman's hand. Traditionally, she can repeatedly pour grain from a small basket held four feet high into a larger one on the ground with just the right touch so loosened hulls will blow away.

At this point the rice is brown, for the tight bran layers are still intact. The next step is critical, for the bran contains valuable vitamins and minerals, yet rice is considered tastier, keeps better, and cooks with less precious fuel if the bran is polished off by milling. In India, and more recently in the production of converted rice in the United States, rice is parboiled before polishing, which allows some of the soluble vitamins and minerals in the bran to permeate the endosperm and add nutritional value to subsequently polished kernels.

The last step in commercial processing is tumbling in glucose or talc to add the sheen that gives this precious grain the beauty of a pearl.

What about the paddy fields, the stalks, the hulls, the bran and the broken kernels? Do they serve only to bring the familiar steaming, fluffy white mounds to the table and nothing more? The answer is of course, no, for by-products are many and rice is a versatile cereal.

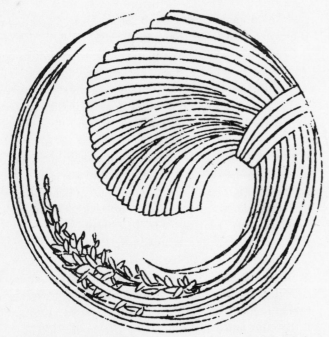

FISH: Frugal rice growers long ago recognized that paddies are ideal for raising fish. Carp, highly prized as food in the Orient, thrive in rice paddies. They help keep the paddies clean, add fertilizer for the plants and are easily caught as they flounder in the mud when the paddies are drained.

STRAW: Rice straw can be baled and fed to cattle or used to bed them down. In China, it is used for mushroom culture. It can also be used for rope, baskets, hats, snowshoes, mats, fuel or making paper (though most "rice" paper like that used for translucent shoji screens comes from mulberry tree fibers and edible rice paper is baked from rice flour). In Japan, where rice is revered second only to Buddha, "tatami" floor mats of honored rice straw symbolize cleanliness, so important to Japanese households that shoes are removed before entering. Rice represents prosperity, so symbolic thick rice straw ropes are tied across thresholds of homes and shrines at New Year time.

HULLS: Because of their bulk and light weight, rice hulls are ideal for packing and insulating and for poultry bedding. They are important as a raw material for cellulose products, manufacturing and as a soil conditioner, often being returned to the growing paddies. Hull ash is useful, too, as a bleaching agent, as filler for bricks, in making soap and other polishing and cleansing agents, as a fine abrasive, as a grease-absorbing compound, and as an ingredient of a matte white glaze for pottery.

BRAN: Rich in vitamins and minerals, rice bran is an ingredient of many manufactured cereals and is available in health food stores for adding to breads, cookies, meat loaves, etc., for extra nourishment. It is a high quality livestock and poultry feed. Oriental women used to wash their skin with pouches of rice bran for the same reason women today rub in various vitamin oils and creams.

Low cholesterol rice oil is extracted from the bran for salads and cooking, especially for deep frying since it retains little odor or taste and is therefore reusable. It has industrial applications in conditioning leather, treating metal and spinning textiles.

POLISH: As rice is polished in successive steps the bran is gradually removed. The final operation removes the last traces of bran along with part of the

endosperm, and this polish, very digestible and rich in B vitamins, is valuable for baby foods and for thickening gravies, sauces and puddings. Rice polish also finds use in soap and button manufacturing.

BROKEN RICE AND BEVERAGES: The kernels that break during harvesting and milling, aesthetically displeasing in a bowl of cooked rice, are by no means wasted. These "screenings" or "second heads" provide food at a reduced price for less particular people and are also a source of starch for flour, glue, paste and face powder. Some are used for cattle feed, but many end up in beverages and vinegar and are known as "brewers rice".

Marco Polo, returning from Peking, wrote of a "liquor which they brew of rice with a quantity of excellent spice in such fashion that it makes better drink than any other kind of wine. It is not only good, but clear and pleasant to the eye. And being very hot stuff it makes one drunk sooner than any other wine". Though he may have been referring to *shemshu*, the Chinese red wine, more probably it was *arrack*, the rice whiskey.

The famous Japanese distilled rice wine, *sake*, originated as a sacred drink for festivals and other special occasions. Made from steamed rice leavened into dextrose and fermented, it is served hot before or after meals. The sediment left over is a good meat and vegetable preservative. *Mirin* is a sweet *sake* used for cooking.

In India there is a potent rice beer called *bakhar*. In the United States and Europe rice is often used for malting and brewing beer.

FLOUR: Rice contains no gluten, so when ground into flour is unsuited when solely used for baking yeast breads, pies and pastries. Rice flour may be substituted in other breads, however, and is excellent for breading or flouring when sautéing or deep-frying. It makes fried chicken crispier and less greasy, and by mixing with water can be made into "paper" for pastry confections or edible inner wrapping for candies. The Vietnamese eat thin, translucent sheets of rice paper with their meals, while the Chinese use it for parchment shrimp or chicken. Rice wafers, very similar to rice paper, are taken at communion in some Christian religions.

THE ALMOST PERFECT FOOD

"Rice is good at all times and for all persons of any age, whose humours are too sharp and much agitated, and for those, who having impaired their strength, stand in need of some food to restore them."

—*"A Treatise of Food"*,
an old English cookbook

Most Americans don't fully appreciate the great versatility and outstanding nutritional value of rice. If they did, annual consumption per person might rise substantially from the present eight-pound average, though surely never reach the 400-pound level found in some Oriental countries. It is in these high-consumption areas that rice most dramatically demonstrates its versatility — in soups, salads, entrees, entree accompaniments, desserts and beverages — in the rice curries of India, rice and lentil combinations of Ceylon, rice and beans (pah jook) of Korea and the rice and coconut of Hawaii, to name but a few of the thousands of delicious rice dishes world travelers encounter.

American scientists, however, in their continued study of this ancient food, are placing rice higher and higher on their scales of nutritional value. In 1973, a Harvard professor of nutrition rated it "excellent" in protein quality — not quantity — along with cheese, fish and meat. Only eggs and human milk were rated "superior". Beef was ranked lower than many other foods in "biological value" because of the saturated fat and cholesterol one consumes along with the protein. The fat content of rice is almost negligible.

Eight of the eleven essential amino acids — building blocks of food proteins the human body metabolizes, or converts, into its own vital proteins — are present in rice, which is considered to have the best, but not the most proteins of any cereal grain. Unpolished rice contains vitamin E and the B-complex vitamins thiamine, riboflavin and niacin, and of course even polished rice is loaded with easily-digested (low fiber content), nonfattening, energy-producing carbohydrates. A meal of rice is non-allergenic and takes only an hour to digest, far less time than most other foods, making it ideal for people with dietary problems. Veterinarians often recommend it as a low-bulk, bland pet food.

Minerals include phosphorus, iron, potassium and calcium, with a low content of sodium — a good combination for those whose doctors prescribe a salt-free diet. Rice contains no gluten, the tough, viscous substance found in wheat.

The saga of man's struggle to overcome beri-beri, like that of his fight against scurvy, is a classic in the annals of medicine. Rice for hundreds of years has been ascribed medicinal virtues. Rice water was prescribed for high fevers. Rice cakes fried in camel fat were used for hemorrhoids, rice flours as a dentifrice, rice stalks for biliousness, and hull ash for treatment of wounds and discharges.

But only in the last century have scientists discovered the role these complex, elusive chemical compounds they now call vitamins play in the life and the health of all animals. Rice really was the almost perfect food until man decided to grind off the bran layers to make it look prettier, store longer, cook faster, and taste better. Millions died of dread beri-beri before it was proved no

virus or bacteria was to blame, but rather a serious dietary deficiency, lack of vitamin B-1, a sufficient amount of which was eliminated from the diets of rice-lovers by polishing. Even today many people suffer because they subsist on polished rice with no supplementary foods that could supply this vitamin. Parboiling to permit it to soak from the bran layers into the kernel is now widely accepted in India, and brown rice is being promoted as a more perfect food in problem areas. Excessive washing or cooking in large amounts of water leaves out more nutriment. People from the Philippines are smart enough to serve the excess rice water to their young children.

There is more to eating than merely ingesting nourishment to survive, more to living than merely surviving. Confucius in 500 B.C. knew this well as he preached the gospel of a virtuous, yet graceful life. He was a stickler for excellence and ceremony at the table and insisted on the pure whiteness of rice in sheer, elegant porcelain bowls as a background for light emerald-green vegetables picked at their succulent zenith, golden brown stir-fried morsels of duck, pork or fish, and deep red jujube dates. "Tung goh shik fahn" (come eat rice with me) is the most gracious greeting in Chinese hospitality. In old China, families kept two crocks of rice, a large one of gleaming, white, polished rice for the family, a smaller one of coarse brown rice for beggars seeking one more day of existence.

Ironically, the beggar got more food value, but then that was all he ate, while the family supplemented its rice with vegetables and meat. To upset a rice bowl accidentally forebode misfortune; to deliberately empty a bowl of rice on the ground was the ultimate insult, a portent of great calamity. One proverb proclaims a dinner without rice is like a pretty girl with only one eye. Unless the Chinese have rice with their meal they feel they haven't eaten. To those who profess to have "hungry pains" a half hour after a Chinese meal they would say, "Maybe you didn't eat enough rice", for rice has a unique capacity to absorb the impact of "heavy" foods and seasonings and may even help one eat more and feel less stuffed.

The thousands of varieties of rice available in markets of the world can be categorized into several groups:

"WILD RICE": *Zizania aquatica*, also known as Indian rice, grows wild in marshes of the Great Lakes region, but is not a true rice. By law it is only harvested by knocking ripe grain into boats, since machines would clean stalks so thoroughly no reseeding would occur. Its unique nutty flavor and texture comes out only after long soaking followed by cooking until tender and serving with butter or preparing as a pilaff. For stuffing wild game it is unsurpassed, but its price can be prohibitive. Brown and white rice combined make a good substitute.

LONG GRAIN: Most popular with the Chinese and with Western connoisseurs, the grains are four to five times as long as they are wide and are translucent. Cooked properly, the grains are distinct and separate and are especially good in salads, curries and gumbos, or in continental cooking as accompaniments for meat, poultry and seafood dishes. Varieties include: Carolina, Patna, Blubonnet, Labelle and Starbonnet.

MEDIUM AND SHORT GRAIN: The length of medium grain kernels is about three times their width; short grain, one-and-a-half to two times. The shorter the grain, the shorter the growing time and the higher the yield. Together with easier milling these factors substantially reduce prices. Shorter grains tend to be more moist, require less water for cooking and cling more to each other. They can be substituted for long grain and are especially suitable for molded dishes such as croquettes, puddings, rice rings, and *sushi*. Puerto Ricans, Koreans and Japanese prefer short or medium grain rice. Medium varieties include: Nato, Nova Vista, Saturn and Calrose; short varieties include Pearl, Colusa and Caloro.

MILLED VS BROWN: All kernels come in polished (milled) or unpolished (brown) form. Both have the inedible hulls removed, but brown rice retains its bran layers containing oils, proteins and vitamins so nourishing to animals and also to insects (who will shun polished rice) and should be refrigerated or kept cool and dry if stored for prolonged periods. Brown rice has a chewier consistency and a slightly nutty flavor, requires more water for cooking and usually twice the cooking time, which can be shortened by prior soaking.

CONVERTED AND PARBOILED: Steam or hot water processing before polishing allows water soluble proteins and vitamins and some of the bran's nut-like flavor and color to permeate the starch endosperm and remain there. Products so labeled may be off-white or tan, though not necessarily very pronounced, and require somewhat more cooking time.

ENRICHED: To replace nutriments removed in polishing, extra vitamins are sometimes added during processing in accordance with government nutritional standards. Labels will so specify.

PRECOOKED OR INSTANT: Especially useful for cooks in a hurry, precooked or instant rice needs only to be heated quickly in boiling water to be ready to serve. Follow package directions.

GLUTINOUS OR SWEET: This short, opaque, off-white grain with more dextrose and some maltose contains no gluten as the name would seem to imply. Very sticky when cooked it is a preferred variety in Laos and is used extensively in Oriental ceremonial foods and specialty dishes.

In Japan, the most important New Year food is mochi, a gelatinous rice cake made by pounding hot, steamed rice into a sticky dough. Mochi Tsuki, the rice pounding ceremony, takes place a few days before the New Year and is a festive occasion at the rural homes that still observe it. Only men, working in pairs, are permitted to wield the big mallets that pulverize the mass of steamed rice in a wooden tub. When finished, the dough is formed into round cakes (symbolizing a mirror, one of the three ancient imperial treasures) which are placed as offerings to the Shinto gods. Mochi also means "to have", symbolizing wealth. The cakes are very practical and can be kept for days and prepared for eating merely by toasting over a fire, so during the New Year season, which is a time of rest for all, "mama san" need not cook.

TOOLS & TIPS

An inveterate rice eater and cook will experiment with the many fascinating pots, utensils and serving dishes available today. Selecting those most suitable is a matter of individual choice, but there are several general considerations.

RICE POTS: Your rice pot may well be the most critical pot in the kitchen if you use it day in and day out. Many cooks prefer a thick, cast aluminum pot with a tight-fitting lid because it retains heat, keeps moisture in and helps prevent scorching if you use the Oriental cooking methods. I find an ordinary stainless steel saucepan with tight lid equally satisfactory when closely watched. If one prefers cooking rice in a pot of boiling water, any large pot without a lid will suffice. Because rice at a boil tends to produce a thick layer of bubbles, the Japanese have designed an ingenious pot with an expanded upper rim extending out from and above the lid to allow the froth to dissipate without overflowing. Adding a tablespoon of oil to the water helps prevent bubbling over when using any type of cooker.

Often there is a crust left in the bottom of the pot. To some, this "stuck" rice is considered a mistake and is thrown away. To me it's a delectable snack. I carefully lift it out with a spatula, spread it with foo guey (Chinese fermented bean curd similar to brie cheese) and savor it as an after dinner treat, a kind of Oriental cheese and crackers. Persians delight in eating their buttery chilau rice crusts, and Indonesians eat their crusts (kerak) with chili. Often rice crusts are made into special tea by pouring boiling water over to extract the flavor, in Japan also green tea, bonito flakes and nori. If crust is not desired use a teflon saucepan.

ELECTRIC RICE COOKERS: These marvelous cookers are experiencing growing popularity among the Chinese and Japanese and throughout the world, for they are portable and foolproof if directions are properly followed. No scorching, no need to watch the pot — just press the button and the cookers will shut themselves off, yet keep the rice warm until serving. They are even suitable for cooking other foods, either as a separate dish or on a special steaming rack above the boiling rice. They may supplement your stove at home or accompany you on your travels to prepare hot food wherever you are.

RICE BALLS: The rice ball is made of aluminum with perforated holes all around. It consists of two halves held by a hinge and clasp. There are two sizes, holding one or two cups raw rice. The rice is placed in the ball and dropped into boiling water for cooking. When rice is done, the ball is lifted out and any excess water drains from the perforations. The rice may be partially cooked in the ball, emptied and added to a dish with other sauces or flavorings for finished cooking. Allow room for the rice to expand within the rice ball; otherwise the rice may not be thoroughly cooked or a sticky ball will result.

PAELLA PANS: The famous Spanish rice dish is named after the pan, *paella*. It is a large, flat, metal pie-shaped pan with a handle on each side. In Spain, these pans can measure up to three feet across! A good size pan to feed six to eight should be approximately 16 inches in diameter. Be sure to measure the width and depth of your oven before purchasing an extra large paella pan. Authentic pans are made of hammered steel and are imported from Spain. The handles should be steel, without any wood or plastic. A paella pan is a frying pan, rice cooker, baking and serving dish combined. It does not have a lid as there is no need to cover when cooking a true paella.

BAMBOO STEAMERS: Bamboo steamers are used for cooking various steamed dishes as well as raw rice wrapped in banana, lotus or bamboo leaves. It is always necessary to soak the rice for several hours (overnight for glutinous rice) before steaming to provide enough moisture. This method takes considerably longer than others. The Indonesians, especially, often use steamers for cooking their rice. The steamer is filled with parboiled rice and placed over a large pot or wok of gently boiling water and the rice is steamed until tender.

RICE BUCKETS: After the rice is cooked, a good Japanese housewife will put the perfectly cooked steaming rice in a "kama", a wooden or lacquer bucket or bowl with a cover, and place this treasured and honored rice to the right of the master at table. The "kama" absorbs excess steam from the rice and at the same time keeps it hot. The wooden paddle to scoop out the rice is left in the covered pot.

PADDLES: These handy hardwood or bamboo paddles from Japan are great for scooping rice into a bowl. Because of their flat surface, any adhering rice can easily be scraped off against the surface of a rice bowl. Though the wooden paddle remains in the "kama", it is never hot to the touch.

RICE BOWLS: Rice bowls come in various sizes and shapes. A well-designed bowl must be comfortable to

hold in the palm of your hand. Filled with warm, wonderful rice, the bowl is always lifted to the lips when the rice is eaten with chopsticks in the Oriental manner. A rice bowl is all purpose: it can be a soup bowl, a bowl for fruit, an ice cream or nut dish, or even a jello mold. Just about anything and everything tastes better in a rice bowl!

The rice grain pattern on china originated during the Ming Period (1368-1644) and was a form of decoration

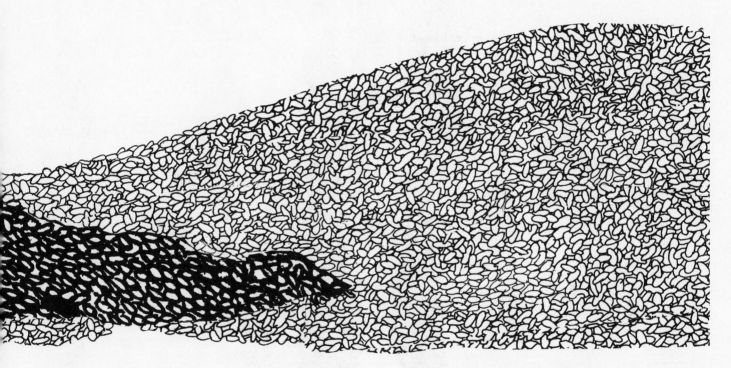

for fine, thin porcelain. Potters applied rice grains with utmost delicacy and deftness into the thin walls of the body while in its air-dried state. Then the piece was glazed and fired. The glaze would fill the burned-out rice piercings, leaving the rice grain pattern absolutely transparent. On blue and white china, especially rice bowls, this pattern remains ever popular even to this day.

CHOPSTICKS: Oriental rice purists claim serving bowls and implements should be wood, ceramic or ivory — never metal. In China, the privileged class felt if ivory showed any stains, the food had been poisoned or was impure. I prefer chopsticks because even a silver fork has a noticeable metallic taste. Chopsticks of various lengths are very useful and simple tools, not only at the table, but also in the kitchen for stir-frying and using as tongs for lifting or turning morsels of food. It is the perfect tool for fluffing cooked rice.

MAKING EVERY KERNEL A KING

Because the moisture content of rice differs with the variety and age of rice, it is difficult to give exact measurements and insure a finished rice that will always turn out the same. Newly-harvested rice, for instance, requires less water when cooking than does old rice. If you cook rice by the Chinese or Japanese method and buy your rice in 50- to 100-pound sacks, you will discover each sack may be slightly different. Experiment and add or delete water to make a perfect cooked rice. When buying in quantity at a considerable savings, store in a dry place away from heat and keep off concrete floors if leaving in the sack. It's best to transfer rice from sack to a bin, plastic or galvanized can or large crock.

Every nationality cherishes its own centuries-old way of preparing rice for the table. Some prefer their rice dry, others moist. As long as rice is cooked thoroughly, eaten and enjoyed it really doesn't matter which method is used. Boiled or steamed rice grains should emerge whole, tender and separate. If the method is done correctly there is no excuse for a dish of rice to be sticky, gummy or mushy.

In the United States the tendency is not to wash rice before cooking. The Oriental and other rice-eating people prefer to wash the raw rice to rid it of excess floury starch which results in a cleaner taste and a fluffier cooked rice.

- **Brown Rice.** Soak brown rice 1 hour before proceeding with any of the preceding methods. Increase cooking time by 15 to 20 minutes. When preparing by the Chinese method, add an additional tablespoon of water per cup of brown rice.

The Chinese Way

Most Chinese prefer long grain white rice for their daily meals. As do most Orientals, they eat more rice with their meals than the amount usually eaten with continental meals.

STEAMED LONG GRAIN (CAROLINA OR PATNA) RICE

When cooking by this method, always cook at least 1 cup of rice. Plan 1/3 cup per person. Wash in a saucepan with a tight-fitting lid by "swishing" around with fingers under cold running water. Pour off water and repeat until water runs clear, about 5 times. Level rice in pan and fill with water to cover by 3/4 inch (about 1 cup rice to 1 1/4 cups water). Use the proper size pot for the amount of raw rice being cooked. The pot should be at least double the amount of rice being cooked to allow enough room for expansion during cooking. Never use a huge pot to cook a small amount of rice or the end product will be more crust than rice. Rice that is washed at least a half hour before cooking will be more uniformly cooked because the kernels have had a soaking period and will thus need less time to cook evenly and thoroughly. To begin cooking the rice, place over *high heat*, uncovered, until water begins to boil. Lower heat to *medium* until all water and bubbles disappear. Cover pot, reduce heat to *lowest simmer* and continue cooking 15 minutes. Fluff up with chopsticks and serve. The crust that forms on the bottom of the pot may be eaten as is, saved for sizzling rice soup, or combined with boiling water to make rice tea. For thicker crust increase cooking time to 45 minutes.

The Japanese Way

STEAMED MEDIUM OR SHORT GRAIN (PEARL) RICE (CAL ROSE AND CALORO)

Called Gohan in Japanese and preferred because when cooked, short grain rice is moist and the kernels tend to stick together. In Japan, a bride is judged by her new family by the quality of her rice cooking. She knows where the rice comes from — its moisture content and age. From November to mid-December and from April to June, she will cook rice with equal amounts of rice and water. From July to September, however, she will use 9 cups of water to 8 cups of rice. By the end of September to the end of October, the proportions will be 8 cups rice to 10 cups water. When the tastier newly-harvested rice arrives in November there are harvest celebrations throughout Japan.

Always cook a minimum of 1 cup rice if cooking by this method. Cook the Chinese way but add ½ inch water above the rice level. Or, drain well-washed rice, place in saucepan with tight-fitting lid, add 2¼ cups cold water for 2 cups rice and soak 30 minutes. Over high heat bring to boil, reduce heat to medium, cover and continue cooking 8 to 10 minutes until water has evaporated. Lower heat to simmer and cook 5 minutes longer. Remove from heat and let rest 5 minutes, covered.

The Indonesian Way

BOILED AND STEAMED RICE

Boil 2 cups rice in 2½ cups water until water is absorbed. Transfer rice to a colander or steamer set over a pan of boiling water. Cover pan and let steam 20 minutes or until tender. If using glutinous rice, (soak 4 hours) increase steaming time by 10 minutes.

Variations:

- Add ½ teaspoon salt and 1 bay leaf when cooking.
- Add ½ teaspoon powdered turmeric when cooking.
- Substitute coconut milk for half or all the water.

Makes 6 cups cooked rice

The Brazilian Way

In many Brazilian homes, a meal would not be complete without a bowl of rice and another of black beans. In heavy pot or skillet with tight-fitting lid, saute 1 onion, chopped, in 2 tablespoons oil until transparent. Add 1 cup peeled, chopped tomatoes, bring to boil and add 1 cup rice. Stir to prevent sticking. When rice begins to absorb the moisture from the tomatoes, add 1¼ cups boiling water, 1 teaspoon salt and ½ teaspoon freshly ground pepper. Do not stir. Bring to boil, cover, lower heat and simmer 25 minutes or until all liquid is absorbed and rice is tender.

Makes 4 cups cooked rice

The Cuban Way

The Cubans finish cooking their rice in the oven. In a 3-quart saucepan with a tight-fitting lid place 2 cups long grain rice, washed and well drained. Cover with 2 quarts boiling water and 1 teaspoon salt. Bring to rapid boil, cover lower heat to a simmer and cook 15 to 20 minutes or until tender. Empty into a colander and rinse with cold water until water runs clear. Spread 2 tablespoons butter on bottom of baking dish, add rice and top with another 2 tablespoons butter. Place a piece of brown paper (paper bag) over rice and bake in preheated 250° 10 to 15 minutes to heat through. Every grain should be fluffy and separate.

Makes 6 cups cooked rice

The American Way

BOILED RICE

Bring a large pot of water to the boil. Salt lightly if desired and gradually add one cup rice, keeping the water boiling constantly. Boil 10 to 12 minutes or until al dente, stirring several times to be sure rice does not stick to bottom of pan. Drain in a colander and run hot running water through to rid rice of any excess starch. Place well drained rice in baking dish. Dot with butter (optional) and keep warm in a 250° oven up to 30 minutes. Fluff up with fork just before serving.

Makes 3 cups cooked rice

The Baked Way

Combine 2 cups white or brown rice, soaked 30 minutes, 4 tablespoons butter, 3 cups boiling water or stock, 1 onion, minced, 1 tablespoon minced ginger root and 1 teaspoon salt. Place in casserole, cover tightly and bake in preheated 325° oven 35 minutes for white rice and 50 minutes for brown. Fluff with fork and serve as accompaniment to eggs, meat or seafood.

Makes 6 cups cooked rice

The Pilaf Way

SAUTÉED AND STEAMED RICE

In heavy skillet or 2-quart saucepan over moderate heat, melt 2 tablespoons butter. Add 1 cup long or medium grain rice and cook and stir until grains are evenly coated. Do not brown. Pour in 2 cups hot chicken stock and, stirring constantly, bring to boil. Cover tightly, reduce heat to simmer and cook 20 minutes, or until tender. Toss with 4 tablespoons butter and fluff well until grains shine with the butter coating. Salt and pepper to taste. Place a towel over the rice to absorb any extra moisture and let stand at room temperature 10 minutes before serving.

Makes 3 cups cooked rice

Variations:

- **Saffron Rice.** For delicate color and subtle fragrance dissolve ⅛ teaspoon saffron threads in hot stock.
- **Spinach Rice.** Add 1 cup finely chopped and well drained chopped cooked spinach to chicken stock with a pinch of nutmeg. Be sure to mix the spinach evenly into the rice. Good with ham or poultry.
- **Tomato Rice.** Add 2 tablespoons tomato paste to stock or reduce stock to 1 cup and combine with 1 cup

seeded, chopped tomatoes. Just before serving toss in 1 tablespoon or more chopped fresh coriander or 1 teaspoon chopped fresh basil (optional).
- **Caraway Rice.** Toss 1 teaspoon caraway seeds into the cooked rice.

Glutinous Rice

Prepare the Chinese way. Soak at least 4 hours before cooking and increase cooking time to 35 minutes.

Converted or Parboiled Rice

Increase cooking time 10 to 15 minutes.

Precooked or Instant Rice

Follow directions of the package.

Cooking yield of rice:

- 1 cup uncooked long, medium or short grain yields approximately 3 cups cooked.
- 1 cup uncooked Converted or Parboiled rice yields approximately 4 cups cooked.
- 1 cup uncooked Precooked or Instant rice yields approximately 2 cups cooked.

Rice Variations

- Cook rice the Japanese way, adding sweet yams, peeled and cut in chunks, when boiling rice.
- Cook rice, adding a pinch of turmeric to liquid. Toss in extra butter and serve with fish for color contrast.
- Cook glutinous rice as directed, allowing crusts to form. Break up crusts and serve the Laotian way with savory hot sauce.
- To 3 cups cooked rice add 2 tablespoons lemon juice and 1 teaspoon grated lemon rind.
- Toss a pat of butter into cooked rice and top with a sprinkling of paprika, chopped chives, crumbled cooked bacon and several dabs of sour cream.
- Into cooked rice toss sautéed mushrooms, sliced water chestnuts, ripe olives, toasted slivered almonds, chopped pine nuts or peanuts, freshly grated Parmesan or Cheddar cheese, chopped green pepper or green onion, chopped ripe tomatoes, cooked fresh peas and/or chopped parsley or chervil. If desired, pack into buttered ring mold and unmold on heated platter.
- For breakfast or mid-day serve cooked rice with milk or cream (hot or cold), honey or sugar, raisins, dates or apricots, nuts or sunflower seeds, or fruits of the season.

TO REHEAT RICE

1. Put rice in a saucepan with a tight-fitting lid. Drizzle a tablespoon of water for every cup of rice over surface, cover and simmer over low heat until heated through.
2. Put rice in heat-proof bowl. Set bowl above 2 inches water in large pot to create steam. Cover pot and steam over medium heat 10 minutes or until hot.
3. Put rice in a sieve or colander and reheat as above.
4. Put rice in an oven-proof dish or pan, drizzle 1 tablespoon of water for each cup of rice over surface. Cover and heat in preheated 300° oven 20 minutes.

SAVING A POT OF BURNT RICE

If you should accidentally scorch a pot of rice when cooking the Chinese way, place a slice of bread and 1 or 2 tablespoons water or a moistened clean cloth napkin on surface of rice, cover, lower heat and cook until tender. The burnt smell will be absorbed.

APPETIZERS & SALADS

Stuffed Celery With Cream Cheese, Rice and Chutney

1 3-ounce package cream cheese, softened
2 tablespoons milk
1 cup cooked rice
3 tablespoons chopped chutney
2 tablespoons minced green pepper
2 tablespoons minced onion
2 tablespoons chopped nuts
Salt and freshly ground pepper to taste
4 celery stalks

Whip cream cheese and milk until light and fold in rice, chutney, green pepper, onion, nuts and salt and pepper. Stuff celery stalks and cut in 2-inch pieces.

Variations:

- Add 2 tablespoons crumbled crisp bacon and ½ teaspoon curry powder
- Omit chutney and nuts. Add ¼ cup each minced ham and chopped olives
- Omit chutney and nuts. Add ½ cup minced clams or cooked shrimp and 2 teaspoons prepared horseradish. After filling sprinkle with sieved hard-cooked egg yolk.

Makes approximately 1½ cups filling

Cold Dolmas

Popular throughout the Middle East from Lebanon to Greece, these dolmas, a delicate blend of spices, lemon and rice stuffed in grape leaves, are eaten cold as an appetizer.

1 1-pound jar grape leaves or
40 to 50 fresh grape leaves
1 cup long grain rice, washed and drained
3 tomatoes, peeled and chopped
1 onion, finely chopped
3 tablespoons finely chopped parsley
3 tablespoons finely chopped mint or
1 tablespoon dried mint
¼ teaspoon each cinnamon and allspice
1 teaspoon salt
½ teaspoon freshly ground pepper
½ cup olive oil
⅓ cup lemon juice
1 teaspoon sugar
3 garlic cloves, halved

Rinse grape leaves with cold water, place in large bowl and pour hot water over to cover. This will rid the leaves of excess salt from brine. Drain and cool. If using fresh leaves, wash, dip in boiling water just until limp and drain and cool. Combine rice, tomatoes, onion, parsley, mint, cinnamon, allspice, salt and pepper. Put 1 heaping teaspoon of rice mixture on flat leaf, spread out vein side up, near stem end. Fold stem over filling, bring the two sides over and roll to remaining edge of leaf to form a small cigar shape. Squeeze lightly in palm of hand and place seam side down on plate until all leaves have been filled. Line a large pot with tight-fitting lid with a layer of unfilled grape leaves. Place filled leaves seam side down closely on top. Combine olive oil, lemon juice and sugar and drizzle over filled leaves. Tuck the garlic halves into the pot and add 1 cup water. Place a plate on top to prevent any shifting during cooking. Cover pot and simmer gently 2 hours, adding additional water if necessary. Cool in pot before removing and serve with egg lemon sauce.

Makes 3 dozen

Variation: Omit the mint and add 2 tablespoons fresh chopped dill and ½ cup currants to filling.

Hot Dolmas: Add ½ pound ground veal, beef or lamb to filling, proceed as directed and serve hot.

Egg Lemon Sauce

4 egg yolks, beaten
⅓ cup lemon juice
1 cup hot chicken or vegetable stock
Salt and white pepper

Gradually beat lemon juice into beaten yolks. In a slow stream pour hot stock into eggs, beating constantly. Pour into top of double boiler set over hot, not boiling, water. Cook and stir 7 minutes or until sauce is thick. Season to taste with salt and pepper. Serve warm or cold.

Makes 1½ cups

Rempejek (Rice/Peanut Crackers)

Indonesian rice/peanut crackers eaten as a snack or served as a side dish.

2 cups rice flour
1 tablespoon cornstarch
1 tablespoon ground almonds
1 small onion, grated
1 garlic clove, minced
1 teaspoon ground coriander
1 cup each coconut milk and water
1 cup chopped raw peanuts or unsalted
  roasted peanuts
Salt and pepper to taste
Vegetable oil for deep frying

Combine the rice flour, cornstarch, almonds, onion, garlic and coriander. Gradually add the coconut milk and water, stirring well to make a smooth batter. Add the peanuts, salt and pepper. Drop by tablespoons into hot oil, a few at a time, and fry until golden. Drain on paper toweling. Repeat until all mixture is used.

Makes approximately 3 dozen

Coconut Milk

Combine 1 cup milk and 1 cup unsweetened grated coconut in a saucepan. Bring just to boil over medium heat. Remove from heat, cool and strain, pressing coconut against sides of sieve to extract all the essence. Reserve coconut for puddings or cakes. Makes 1 cup milk. Canned coconut milk may be purchased in most markets.

Risotto and Cheese Balls

Leftover risotto may be used for these delectable and creamy fried rice balls.

2 cups risotto (or any cooked rice)
2 eggs, beaten
¼ pound mozzarella or Monterey Jack cheese, cut in ½-inch cubes
1 cup fine bread crumbs
Vegetable oil for deep frying

Combine rice and eggs. Take a tablespoon of mixture, place in palm of hand, put a cheese cube in center and top with another tablespoon of rice mixture. Form into 1¼-inch balls and roll in bread crumbs. Repeat until all rice balls are made. Chill for 30 minutes. Heat oil to 375° and fry the balls, a few at a time, for 3 minutes or until golden brown. Drain on paper toweling and serve immediately as appetizers. May be kept warm in a 350° oven for 10 minutes until all balls have been fried.

Makes 1½ to 2 dozen balls

Law Bak Go (Turnip Pudding)

A Chinese steamed pudding. Serve as one of many "dim sum" dishes along with a favorite tea, or as an afternoon snack.

1 pound Chinese turnips or daikon, peeled and sliced
2 cups rice flour
¾ cup cold water
1 teaspoon salt
6 tablespoons peanut oil
¼ pound lean pork butt, ground
2 tablespoons tiny dried shrimp
1 teaspoon soy sauce
2 tablespoons chopped green onions
1 tablespoon toasted sesame seeds

Place turnips in a pot with water to cover and cook 30 minutes or until soft. Drain and mash to a pulp while hot. Add the rice flour, water, salt and 4 tablespoons of the oil, blend well and set aside. Sauté pork and shrimp in remaining 2 tablespoons oil for 5 minutes. Add the soy and combine with the turnip and rice mixture. Pour into an oiled square pan or pie plate. Cover pan with a piece of wax paper to catch steam drippings during cooking. Place pan in a large steamer or wok with at least 2 inches of water and cover with lid. Steam over medium heat 1½ hours or until pudding is set and a knife inserted in center comes out clean. Check the water from time to time, adding more boiling water if needed. When set, take out of steamer and sprinkle green onions and sesame seeds on top. Let cool to lukewarm before cutting into 2-inch squares or diamonds. May be eaten warm or at room temperature.

Serves 6

28

Edible Rice Paper Chicken

Approximately 12 sheets edible rice paper*

**1 pound chicken breast, boned and cut in
1-inch pieces**
2 tablespoons soy sauce
2 tablespoons whiskey or gin
½ teaspoon each sugar and Five-spice powder
1 teaspoon minced ginger root
2 garlic cloves, minced
**1 tablespoon each chopped fresh coriander and
green onions**
½ teaspoon Oriental sesame oil
1 tablespoon peanut oil
Vegetable oil for deep frying

Combine all ingredients except vegetable oil. Let
stand at room temperature 1 hour. Drain. Cut rice
paper into 4-inch squares, using a double layer if paper
is very thin. The paper absorbs quickly, so work with
only a few squares at a time, deep frying as they are
filled. Place a little of the chicken above one corner of
paper. Fold envelope-style, tucking in sides to enclose.
Fry immediately in hot oil 3 to 5 minutes until chicken is
just cooked. Remove and drain. Serve hot. Continue
until filling is used.

Makes 2 dozen

*Available through some bakeries or Oriental stores.
Parchment paper may be substituted, but is not edible.*

Pearl Rice Meatballs

These tasty meatballs, rolled in rice and then
steamed, are served as appetizers or a dish for "dim
sum" tea lunch.

¾ cup glutinous rice
½ pound ground pork butt
¼ pound raw shrimp, shelled, deveined and minced
½ cup water chestnuts, minced
**3 dried forest mushrooms, soaked to soften and
minced (optional)**
1 tablespoon each soy sauce, sherry and water
2 tablespoons cornstarch
½ teaspoon Oriental sesame oil
Salt and white pepper to taste
Vegetable oil

Soak the glutinous rice in water for 30 minutes.
Drain. Combine and mix well all remaining ingredients
except oil. Form into 1-inch balls and roll in rice to coat,
pressing rice well into balls. Place not touching on oiled
pie pans. On rack over water in a large pot with tight
fitting lid or in bamboo steamer set over water in wok,
steam 1 hour. Serve hot.

Makes 3 dozen

Variation: Omit pork and shrimp. Add ¾ pound lean
ground beef, lamb or veal, ½ teaspoon minced ginger
root and 2 tablespoons chopped green onions.

Guon Fun

Rolled rice noodle sheets filled with crisp vegetables and meat. Eaten at room temperature for an appetizer.

1 recipe Home-made Fun (page 91)
1 pound bean sprouts, parboiled 2 minutes and well drained
½ pound Chinese barbecued pork or cooked ham, cut in long shreds
1 recipe egg garni using 3 eggs (page 75)
2 stalks green onions, cut in thin lengthwise strips
Coriander sprigs
Red pickled ginger, thinly sliced* (optional)

Oriental sesame oil
Salt and pepper
Soy or oyster sauce for dipping

Place a round noodle skin on a piece of wax paper. Beginning 1 inch above center, place a row of bean sprouts, meat, egg garni, green onions, coriander sprigs and optional ginger. Season sparingly with sesame oil, salt and pepper. Do not fill too generously or sheet will be difficult to roll tightly. Roll filled sheet like a jelly roll and place on plate seam side down. Continue with remaining sheets and filling. To serve, cut into 1-inch lengths and stand up cut side. Accompany with soy or oyster sauce for dipping.

Serves 8 to 10

Sushi Rice (Vinegared Rice)

Sushi rice is the foundation for a multitude of fillings and toppings for these Japanese "rolled open-face sandwiches".

2 cups short grain pearl rice
2½ cups cold water
1 2-inch square kombu (optional)

Wash the rice in several changes of cold water until water runs clear. Drain and put into a 2-quart saucepan with a tight-fitting lid. Add the water and allow the rice to soak for 30 minutes. Add the kombu to the pot and over high heat bring to a rapid boil. Reduce heat to medium, cover and continue cooking for about 8 to 10 minutes until all water has evaporated. Lower heat to a simmer and cook 5 minutes longer. Remove from heat

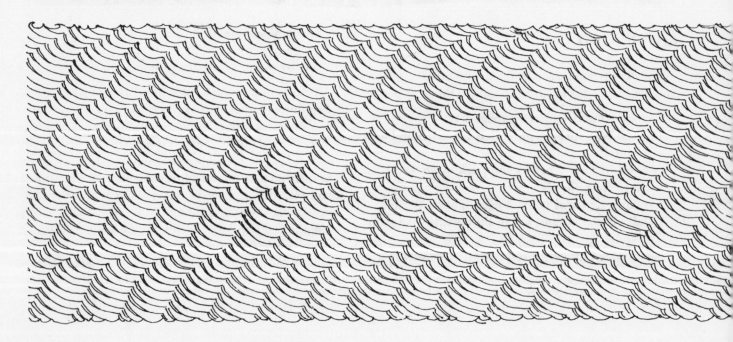

and let rest 5 minutes, COVERED. Discard Kombu and transfer rice to a wooden bowl (this helps to cool the rice and evaporate excess moisture faster than metal, ceramic, etc.). Immediately pour on the vinegar dressing, mixing well with chopsticks or fork. Begin fanning (cardboard works fine) the rice to cool it as quickly as possible. Rice is ready to use when cooled to room temperature.

Makes 6 cups

Vinegar Dressing for Sushi

¼ cup rice vinegar or 3 tablespoons mild white vinegar
3 tablespoons sugar
2 teaspoons salt
1 tablespoon Mirin (sweet sake) or dry sherry

Combine all ingredients thoroughly.

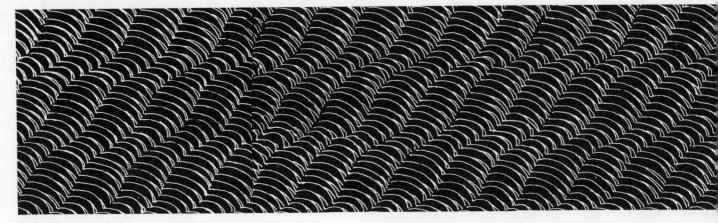

Maki Sushi

Vinegared rice, vegetables and fish rolled in seaweed. Serve as appetizers or for a light lunch. Great picnic food!

1 recipe sushi rice (preceeding)
1 cup dashi stock*
12 sprigs watercress, trimmed of large stems, or
12 spinach leaves
6 sheets dried laver (nori)**
1 2¼-ounce flat-tin canned eel, drained, or
1 6½-ounce can tuna, drained
3 tablespoons thinly sliced pickled red ginger**

Have ready the sushi rice dressed with vinegar. Heat the dashi stock to boiling. Add watercress or spinach and parboil for 1 minute just to wilt. Drain, cool and squeeze out excess water from greens. Set aside. Hold 1 sheet of laver at a time over a candle or low flame to intensify the flavor and toast lightly. Place a sheet of laver on a bamboo mat or cloth napkin. Divide sushi, fish and ginger into 6 portions. Spread one portion of rice on the laver to almost cover the entire surface, leaving a 1-inch border uncovered on one horizontal edge. Along the center, horizontally in rows, place 2 sprigs of watercress or 2 spinach leaves and a portion of the fish and ginger. Grasp the bamboo mat along with the laver and filling and roll tightly like a jelly roll, tucking in the open ends. Unroll the mat and cut filled laver into 1-inch rounds.

Makes 6 rolls

Note: Any combination of cooked fish or seafood may be used as well as egg garni (page 75), sautéed mushrooms, bamboo shoots and kanpyo (gourd shavings).

Dashi stock is made from kombu, a seaweed, and katsuobushi, dried bonito fish. Use the instant variety available in Japanese markets, Dashi No Moto.

**Available in Oriental markets.*

Nigiri Sushi

Oblong rice mounds topped with a variety of raw fish, cooked shrimp, abalone slices, squid or fish cakes.

1 recipe sushi rice (preceeding)
Any combination of seafood for toppings
Wasabi*, mixed with water to form a thick paste
Japanese soy sauce for dipping

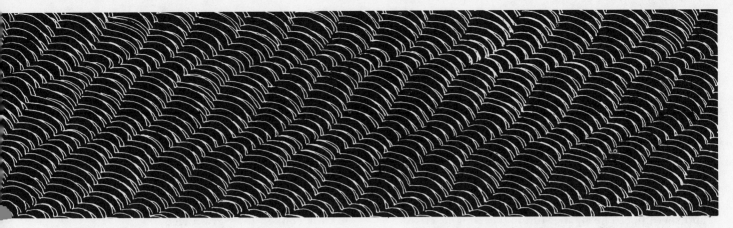

Wetting hands with cold water to prevent sticking, take a heaping tablespoon of sushi rice in the palm of your hand and form and press into an oblong shape. Put a tiny dab of wasabi on each oblong and place a piece of seafood on top to cover completely. Arrange on serving platter and serve with soy sauce.

Green horseradish powder available in Oriental markets.

Hom Goh

Hom Goh may be served as an appetizer or one of several "dim sum" (touch the heart) dishes for a tea lunch. Often prepared during the Chinese New Year for family and friends, these tasty, meat-filled pastries are enjoyed by all.

3 cups glutinous rice flour (sweet rice flour)
Approximately 1 cup hot water
½ pound lean ground pork
¼ pound raw shrimp, shelled, deveined and minced
½ cup minced raw ham
8 dried forest mushrooms, soaked to soften and
minced

2 tablespoons minced green onions
6 water chestnuts, minced
1 tablespoon soy sauce
½ teaspoon Oriental sesame oil
½ teaspoon salt
Vegetable oil for frying

Combine glutinous rice flour with hot water, stirring constantly with a pair of chopsticks or a fork. When lukewarm, place on lightly floured board and knead with hands until smooth and elastic. Let stand 20 minutes. To make the filling, heat 2 tablespoons of the oil in a wok or skillet until very hot. Stir fry the pork until it loses its pinkness. Add remaining ingredients except frying oil and stir fry another 3 minutes. Remove from heat and cool. Form dough into a rope 1 inch in diameter and pinch off balls about the size of a walnut. Flatten with hands or rolling pin into 3-inch circles. Place 1 tablespoon of filling in center, fold over into half moon shape and pinch edges together. Repeat with rest of dough and filling. Heat oil to 230° and fry the hom goh until lightly golden. Drain on paper toweling and serve hot. May be prepared ahead of time and heated 10 minutes in 350° oven.

Makes approximately 2½ dozen

Polynesian Fried Shrimp

1 pound raw shrimp, shelled and deveined
½ cup rice flour or cornstarch
Salt and white pepper to taste
1 egg, beaten
1 cup crumbled py mei fun*
Vegetable oil for deep frying

Dust shrimp with rice flour, salt and pepper, dip in beaten egg and roll in py mei fun. Deep fry a few at a time in hot oil until golden. Drain on paper toweling and serve hot. The py mei fun makes a light, crispy coating. Firm fish fillet or oysters may be prepared in the same way.

Variation: Blend 1 minced garlic clove, ½ teaspoon minced ginger root and ½ teaspoon Oriental sesame oil with the shrimp before dusting with flour.

*Rice vermicelli available in Oriental markets. Also known as behon and mei fun.

Yoghurt and Fruit Rice Salad

3 cups cold cooked rice
2 sweet, juicy oranges, peeled and cut in
small chunks
2 apples, peeled and diced
1 cup diced pineapple
½ cup each shredded coconut and raisins
1 cup yoghurt
1 tablespoon chopped fresh mint or
1 teaspoon dry mint
2 tablespoons lemon juice
Lettuce leaves
½ cup chopped roasted peanuts

Combine the rice, oranges, apples, pineapple, coconut and raisins. Mix the yoghurt with the mint and lemon juice and toss lightly with the rice and fruits. Mound on lettuce leaves and sprinkle with chopped peanuts.

Serves 6

Salad de Riz Nicoise

The French are fond of rice salads and often serve them as a first course.

4 cups cooked long grain rice
2 stalks celery, diced
1 cup peeled, seeded and diced cucumber
½ cup sliced radishes
3 tablespoons sliced olives
1 1¼-ounce tin anchovies
3 tomatoes, cut in wedges
3 hard-cooked eggs, quartered
Lettuce leaves
Watercress sprigs

Mayonnaise-sour cream dressing:

½ cup mayonnaise
½ cup sour cream
1 teaspoon prepared horseradish
1 teaspoon paprika
½ teaspoon dry mustard
Salt and freshly ground pepper to taste

Prepare and chill rice and vegetables. Combine all ingredients for the dressing and blend well. Toss the rice, cucumber, radishes and olives with dressing. Line a salad bowl with lettuce leaves and mound rice mixture in it. Garnish with anchovies, tomato wedges, egg quarters and watercress.

Serves 6

Chinese Chicken Salad

A blend of shredded chicken, peanuts and sesame seeds tossed with light, translucent fried rice vermicelli.

2 whole chicken breasts, split
1 tablespoon each soy sauce and sherry
3 tablespoons hoisin sauce*
3 tablespoons corn oil
1 teaspoon Oriental sesame oil
½ cup chopped roasted peanuts
2 tablespoons toasted sesame seeds
¼ teaspoon Five-spice powder
¼ cup lemon juice
¼ cup slivered green onions
¼ pound py mei fun (rice vermicelli)*, fried
Vegetable oil for frying
Lettuce leaves
4 sprigs fresh coriander

Marinate the chicken with soy, sherry and 1 tablespoon of the hoisin sauce for 30 minutes. Brown well on all sides in corn oil. Be sure to cook thoroughly. When cool enough to handle, shred the chicken meat with fingers and, if desired, thinly slice the skin. Combine the chicken with the sesame oil, peanuts, sesame seeds, Five-spice powder, lemon juice, remaining 2 tablespoons hoisin and green onions. Toss lightly with fried py mei fun and place on lettuce leaves. Garnish with coriander sprigs. Serve lukewarm or at room temperature.

Always combine the pei mei fun with the chicken at the last minute to retain their crispness.

Serves 6

To Fry Pei Mei Fun

Heat at least ½ inch vegetable oil in skillet. When hot, add a small handful of py mei fun at a time. They will immediately expand greatly. Turn over if not completely immersed in oil. Lift out and drain on paper or rack. Repeat until all are fried. They need only be very lightly golden.

Available in Oriental markets.

Rice and Beet Salad

2 cups cooked rice
2 cups diced cooked beets
3 tablespoons minced green onions
½ cup vinaigrette dressing
1 cup mayonnaise-sour cream dressing (page 35)
Lettuce or spinach leaves
Salt and freshly ground pepper
2 cups of any combination of following:

- cooked peas, green beans, asparagus tips, broccoli or cauliflower
- diced apple or pear, or seedless grapes
- cooked diced chicken or turkey, beef or pork
- cooked shrimp, lobster or crab, tuna, salmon or anchovies

Combine rice, beets and green onions with vinaigrette dressing. Chill thoroughly. Just before serving toss with mayonnaise-sour cream dressing and season to taste with salt and pepper. The combination of vegetables, fruits and seafood may either be tossed into the salad or arranged as a garnish around the rice and beet mixture. Line a salad bowl with the lettuce or spinach leaves and mound salad and garnishes as desired.

Serves 6

SOUPS

Old Fashioned Chicken and Rice Soup

1 cup rice
1 plump young hen, about 4 pounds
1 stalk celery with leaves
2 carrots
1 onion
1 bay leaf
2 sprigs parsley, chopped
Salt and freshly ground pepper

In large soup pot combine chicken, celery, carrots, onion, bay leaf and 3 quarts water. Bring to boil, skim off any scum that rises to surface, cover and simmer 1½ hours. Remove chicken, strain broth and return to pot. Add the rice and continue cooking, uncovered, over medium heat for 30 minutes. Add the parsley and salt and pepper to taste. The chicken may be cut up and served separately or it may be boned, diced, added to the soup and reheated.

Serves 6 to 8

Variation: Chicken, Vegetable and Rice Soup. Dice the vegetables. 10 minutes before soup is done add 2 tomatoes, peeled and diced, and 1 cup fresh peas.

Fresh Tomato and Rice Soup

½ cup rice
2 pounds fresh, ripe tomatoes, peeled and cut up
1 quart stock or water
1 teaspoon sugar
1 onion, chopped
3 tablespoons butter
¼ cup chopped fresh sweet basil
Salt and pepper to taste
Freshly grated Parmesan cheese

Combine tomatoes, stock or water and sugar. Bring to boil and simmer 30 minutes. Strain and mash pulp through a sieve. Return to soup pot. In a skillet, sauté the onion in the butter until transparent. Add to soup with the rice, basil, salt and pepper. Bring to boil, lower heat to simmer and continue cooking 20 minutes or until rice is tender. Serve with a sprinkling of Parmesan cheese.

Serves 6 to 8

Variation: Cream of Tomato-Rice Soup. Use 3 cups chicken stock and 1 cup evaporated milk in place of the 4 cups stock or water.

Vegetable Soup with Rice Dumplings

This is a hearty and satisfying "meal-in-a-bowl" soup.

1 good-size, meaty beef or veal knuckle soup bone
1 pound soup meat, preferably from the beef leg,
   cut in small chunks
1 onion, diced
2 stalks celery and tops, diced
2 sprigs parsley, chopped
3 carrots, scraped and diced
3 turnips, scraped and diced
3 beets, scraped and diced
1 bay leaf
Salt and freshly ground pepper

Put the soup bone in a large soup pot with 3 quarts water. Bring to boil and skim off any scum that rises to surface. Add meat, cover and boil gently 1 hour. Add onion, celery, parsley, carrots, turnips, beets and bay leaf. Bring back to boil, lower heat and simmer 1 hour longer. Remove bone and season to taste with salt and pepper. Make rice dumplings and drop by tablespoons onto vegetables in the hot soup. Dip spoon in broth with every drop to prevent sticking so the dumplings will slide off easily. Cover and simmer, without peeking, 15 minutes.

Serves 6 to 8

Rice Dumplings

1 cup cooked rice
1½ cups unbleached flour
2 teaspoons baking powder
1 teaspoon salt
3 tablespoons minced parsley
1 egg, beaten
½ cup milk
2 tablespoons melted butter

Sift flour, baking powder and salt together and stir in parsley. Combine the egg, milk, butter and rice and blend well into flour mixture. Mixture will be stiff and a bit sticky. Drop by tablespoons onto hot soup or stews.

Apple, Curry and Rice Soup

1½ cups cooked rice
2 onions, chopped
2 tablespoons butter
3 apples, peeled and diced
2 tablespoons flour
2 tablespoons curry powder
2 quarts chicken stock
2 tablespoons lemon juice
1 cup or more diced, cooked chicken (optional)
Salt and fresh ground pepper

Sauté the onions in butter until golden. Stir in apples, flour, curry powder and stock. Bring to boil and simmer 1 hour. Add the rice, and lemon juice and optional chicken. Heat through. Season with salt and pepper to taste. Serve hot. Pass a bowl of yoghurt or sour cream.

Serves 6 to 8

Cream of Almond and Rice Soup

½ cup short grain rice
¾ cup blanched almonds, pounded in mortar with
6 bitter almonds or
1 teaspoon almond extract
1 onion, chopped
2 stalks celery, chopped
3 cups milk or half-and-half
2 tablespoons butter
Salt and freshly ground pepper
Freshly grated nutmeg

Combine rice, almonds, onion and celery in soup pot with 1 quart water. Bring to boil and simmer 45 minutes. Put through a sieve and return to the pot. Add the milk and bring to gentle boil. Add butter and season to taste with salt and pepper. Serve in individual bowls with a grating of nutmeg on top.

Serves 6 to 8

Seafood Gumbo

An all-time favorite in the Deep South where shrimps, crabs, oysters and fresh fish are plentiful, always served with a mound of hot, fluffy rice.

1 pound raw shrimp
Fish heads and scraps
1 bay leaf
2 whole allspice
1 onion, chopped
2 garlic cloves, minced
3 tablespoons butter
2 cups peeled and shopped tomatoes
1 pound okra, sliced
1 pound fresh firm white fish, cut in 1-inch chunks
1 pint shucked oysters with their liquor
2 teaspoons salt
½ teaspoon freshly ground pepper
Pinch cayenne pepper
Cooked rice

Shell and devein shrimp. Set shrimp aside and in large pot combine shells, fish head and scraps, bay leaf, allspice and 2 quarts water. Bring to boil and simmer 45 minutes. Strain, discarding shells and fish scraps. Return broth to pot. Sauté onion and garlic in butter for 5 minutes. Add to the soup pot with the tomatoes and okra. Bring to boil and simmer 30 minutes. Add reserved shrimp, fish, oysters and their liquor, salt, pepper and cayenne. Continue cooking 15 minutes longer. Ladle into large individual soup plates with a heaping spoonful of cooked rice.

Serves 6

40

Avgolemono

A tangy, golden soup from Greece.

½ cup rice
2 quarts chicken stock
4 eggs, separated
½ cup lemon juice
Salt and freshly ground pepper to taste
1 lemon, thinly sliced, for garnish

Combine chicken stock and rice, bring to boil, cover and simmer 20 minutes. Remove from heat. In a bowl, beat egg whites until stiff, add egg yolks one at a time and continue beating, slowly adding the lemon juice to the egg mixture. Then gradually add 1 cup of hot soup to egg mixture, beating constantly. Add another cup of soup and continue beating. Return to soup and reheat, stirring constantly. *Do not boil.* Season with salt and pepper and remove from heat as soon as soup is thoroughly heated. Serve immediately in soup bowls with garnish of lemon slices.

Serves 6

Variations:
- Add 1 cup chopped celery and 1 cup diced cooked chicken with the rice.
- Add 1 onion, chopped, and 2 zucchinis, chopped, with rice. Omit lemon slices.

Minestrone Alla Milanese

From the rice-growing region of Lombardy comes this hearty vegetable soup.

1 cup rice
½ cup small white beans, soaked overnight in water to cover and drained
¼ pound salt pork, diced
1 garlic clove, minced
2 sprigs parsley, chopped
1 onion, chopped
¼ pound bacon, diced
3 carrots, scraped and diced
2 stalks celery and tops, diced
2 tomatoes, peeled and chopped
2 potatoes, peeled and diced
3 zucchini, thinly sliced
1 small cabbage, shredded
2 cups fresh peas
1 tablespoon finely minced fresh basil
Salt and freshly ground pepper
Freshly grated Parmesan cheese

Boil beans in water to cover 30 minutes. Drain and set aside. Sauté salt pork in its own fat with the garlic, parsley and onion until soft. Transfer to large soup pot and add 3 quarts water, beans, bacon and all the vegetables except the zucchini, cabbage and peas. Bring to boil, skim off any scum that rises to surface and cook on low heat 1½ hours. Add cabbage, peas and rice and cook for another 20 minutes, or until rice is tender. Add basil and salt and pepper to taste. Serve hot or cold with grated Parmesan cheese.

Serves 8

Mulligatawny

Originating in India, mulligatawny means "pepper and water". It has come a long way from its humble beginning to a tasty and intriguing soup. It can be a meal in a bowl.

1½ pounds mutton or lamb stew, cut in ½-inch dice
4 tablespoons butter or drippings
1 onion, minced
1 tablespoon curry powder
2 tablespoons flour
⅓ cup lentils (optional)
2 tart apples, peeled and diced
1 green pepper, seeded and diced
1 stalk celery, diced
2 carrots, scraped and diced
1 teaspoon sugar
½ teaspoon ground mace
¼ teaspoon ground cloves
Salt and freshly ground pepper to taste
1 cup coconut milk (page 27)
Freshly cooked rice

Brown meat in 2 tablespoons of the butter or drippings. Transfer to large soup pot and add 2 quarts water. Bring to boil, cover and simmer 1 hour. Heat remaining butter in skillet and sauté onion until transparent. Add curry powder and flour and cook and stir for 2 minutes. Combine with the meat and stock and add remaining ingredients EXCEPT coconut milk and rice. Continue simmering the soup for 45 minutes. Just before serving heat the coconut milk almost to the boil and stir into soup. Ladle soup into large individual soup bowls. Serve with freshly cooked rice for everyone to add to his own soup.

Serves 6

Variation: Chicken or fish may be substituted for the meat. Do not brown. Cover with water, cook until tender, remove, bone and return to soup pot just before serving.

Sizzling Rice Soup

The "rice crusts" which form on the bottom of a pot when cooking rice the Chinese way are essential in preparing sizzling rice soup. This is a grand soup to present at the table, for it is a delight to hear the hot rice sizzle when added to the broth.

2 rice crusts, about 6 inches in diameter, cut into
    1½-inch pieces
6 cups chicken stock
1 slice ginger root
1 cup sliced button mushrooms or
6 dried forest mushrooms, soaked to soften and
    sliced thinly
½ cup sliced water chestnuts
¾ cup julienne-cut bamboo shoots
½ pound Napa cabbage or iceberg lettuce, shredded
½ pound chicken breast meat, sliced in julienne or
    1 cup cooked crab meat
1 cup fresh bean curd, cut in ½-inch dice
½ teaspoon Oriental sesame oil
Pinch cayenne pepper (optional)
Salt to taste
Peanut oil for deep frying

Bring the stock, ginger root and mushrooms to a boil and simmer 20 minutes. Discard ginger. Add water chestnuts, bamboo shoots, cabbage and chicken. Return to boil and add bean curd, sesame oil, cayenne and salt. Keep hot. Just before serving heat the oil and deep fry the rice crusts until golden. Have the soup hot and ready at the table so that when the hot fried crusts are added to the soup it will sizzle. *Serve immediately.*
Serves 6 to 8

HINT: Save the crusts from the rice pot when cooking rice the Chinese way (page 20). Freeze or refrigerate until enough crusts have been accumulated. The thicker the rice crust, the better.

Rice And Yoghurt Soup

½ cup rice
4 cups chicken stock
3 tablespoons butter
2 onions, chopped
¼ cup each chopped fresh mint and parsley
Salt and freshly ground pepper to taste
2 cups yoghurt, room temperature

Bring stock just to boil, add rice and cook over medium heat 15 minutes. Heat the butter in a skillet and saute onions until transparent. Add to the stock with mint, parsley, salt and pepper. Simmer 20 minutes. Remove from heat and let stand several minutes. Blend the yoghurt with ½ cup of the hot stock and gradually stir into remaining stock.

Serves 4 to 6

Portuguese Friday Soup

1 cup cooked rice
¼ cup olive oil
1 onion, sliced
2 garlic cloves, minced
1 pound cod or other firm salt water fish,
   thickly sliced
1 slice bread, crusts removed and diced
1 cup peeled, seeded and diced tomatoes
1 small cauliflower, separated into small flowerets
8 cups fish stock or water
½ teaspoon each crushed basil and oregano leaves
Dash of cayenne pepper
Salt and freshly ground pepper to taste
Chopped fresh coriander

Heat oil in large saucepan and saute the onion and garlic 2 minutes. Add the cod and brown lightly on all sides for 5 minutes. Stir in bread, tomatoes, cauliflower, stock and herbs and seasonings. Bring just to boil and simmer 20 minutes. Blend in rice and adjust seasonings. Serve in individual soup or stew bowls and garnish with coriander.

Serves 6

Congee

Called "jook" by the Chinese, congee is a thick creamy-textured soup which can be varied to utilize any extra chicken, duck, turkey, pork, beef, fish bones or meat on hand. It is served as a warming and easily digested breakfast or as a midnight snack.

1 cup rice (some prefer using 2 parts long grain rice to 1 part glutinous rice)
3 quarts water or stock
1 slice ginger root, minced
1 piece dried tangerine peel*, soaked to soften and minced
2 tablespoons minced choong toy (preserved turnip)*
Salt to taste
Chopped green onions and fresh coriander

In soup pot combine rice, stock, ginger root, tangerine peel and choong toy. Bring to boil and simmer 1½ hours or more until rice has completely broken up, adding water or stock if needed and stirring occasionally to prevent sticking. Soup should be thick and smooth. Season with salt and garnish with green onions and coriander.

Serves 6

Variations:
- Add 1 or more cups cooked chicken, pork, ham, beef, etc. just before serving and heat through.
- Use any leftover bones from cooked or raw poultry, meat or fish to make stock. Strain stock before returning to pot with rice.
- Add dried forest mushrooms, soaked to soften and sliced.
- Beat 2 eggs with 2 tablespoons water and stir into soup when ready to serve.

- For the last half hour of cooking add dried bean curd sheets* that have been soaked to soften and cut into 2-inch pieces.
- Add ½ cup tiny dried shrimp*.
- Combine ½ pound ground pork, 1 tablespoon cornstarch, 2 teaspoons soy sauce and ½ teaspoon Oriental sesame oil. Form into balls or drop by small spoonfuls into hot soup last 20 minutes of cooking.

*Available in Oriental markets.

Yuon (Chinese Glutinous Rice Dumpling)

There's an old Chinese saying: "cold, cold, cold; eat a bowl of yuon". The turnips and glutinous rice balls, rich in vitamin B, are bound to keep you warm and hearty. This is a winter meal served morning, noon or night, but never as a soup course. Yuon is traditionally served during the New Year holidays.

2 quarts chicken or pork stock
1½ pounds Chinese or Japanese turnips, scraped and cut into julienne
8 dried forest mushrooms, soaked to soften and thinly sliced
2 tablespoons tiny dried shrimp*
1 piece dried tangerine peel*, soaked to soften and minced
1 slice ginger root
½ pound lean ground pork
1 tablespoon cornstarch
½ teaspoon Oriental sesame oil
2 teaspoons soy sauce
3 lop chiang (Chinese sausage), sliced
Salt to taste
Chopped green onions and fresh coriander

Combine stock, turnips, mushrooms, dried shrimp, tangerine peel and ginger root. Bring to boil and simmer 45 minutes. Combine the pork, cornstarch, sesame oil and soy. Drop by small spoonfuls into the hot stock along with the lop chiang and continue cooking another 15 minutes. Season with salt to taste. In individual bowls ladle soup over the rice dumplings and serve hot with garnish of chopped green onions and coriander.

Serves 6

Glutinous Rice Dumplings

1½ cups glutinous rice flour
½ teaspoon salt
½ cup or more boiling water

Put rice flour and salt in mixing bowl and, stirring constantly with chopsticks or fork, gradually add ½ cup boiling water to form a stiff dough. When cool enough to handle, knead the dough on a floured board for 10 minutes. Roll into ropes ½ inch in diameter and pinch off ½-inch pieces. Roll in palms of hand to form balls. Sprinkle flour on balls to prevent sticking. Repeat until all balls are made. Balls may be made ahead of time. Drop rice balls into briskly boiling water. When they rise to the surface and are puffed, lift out with a strainer, place in individual bowls and pour hot soup over.

Zoni

In Japan, this rice cake soup is traditionally served on New Year's Day.

1 quart dashi stock*
¼ pound fresh spinach or Napa cabbage
1 carrot, scraped and sliced

2 taro roots, peeled and sliced
4 dried forest mushrooms, soaked to soften and cut in quarters
8 mochi**
8 raw prawns, shelled and deveined
1 tablespoon Japanese soy sauce
1 teaspoon salt
8 strips lemon peel

Heat the stock and blanch the spinach or cabbage for 1 minute. Remove and drain. Let cool and cut in 1-inch lengths. Cook the carrot and taro in the dashi for 15 minutes or until tender. Remove from stock and set aside. Add the mushrooms and simmer 15 minutes. Heat mochi under grill but do not brown. Set aside. Add prawns, soy and salt to dashi, bring back to boil and cook 3 minutes. Assemble 8 soup bowls and divide reserved vegetables and mochi in bowls. Ladle the mushrooms, prawns, and stock over vegetables. Garnish with lemon peel.

Serves 8

Dashi stock is made from kombu, a seaweed, and Matsuobushi, dried bonito fish. Use the instant variety available in Japanese food stores, Dashi No Moto.
**Prepared glutinous rice cakes available in Japanese food stores.*

MAIN DISHES

Rice Souffle

1 cup cooked rice
4 tablespoons butter
¼ cup flour
1 cup milk
4 large eggs, separated
½ teaspoon salt

In a saucepan melt butter over medium heat. Add flour, blending well to form a smooth and thick roux. Gradually add milk, stirring constantly until sauce is smooth. Beat egg yolks with salt and add 2 to 3 tablespoons of the white sauce. Return to rest of sauce, stir well and add rice, blending thoroughly. Remove from heat. Beat egg whites until stiff and fold into rice mixture. Pour into ungreased 2-quart casserole and bake in a preheated 325° oven 1 hour. Serve immediately.

Serves 4

Variations:

- Cheese Rice Souffle: Add 1 cup grated Swiss or Cheddar cheese and 1 teaspoon dry mustard to sauce after adding milk. Be sure to stir well to prevent scorching.
- Onion Rice Souffle: Sauté 2 cups chopped onions in the butter until transparent. Then add flour.
- Herbed Rice Soufflé: Add 1 tablespoon minced parsley, 2 teaspoons chopped chives and ½ teaspoon tarragon when adding rice to sauce.

Spinach and Rice Croquettes

2 cups cooked rice
2 cups cooked chopped spinach, well drained
3 eggs, beaten
¼ cup freshly grated Parmesan cheese
½ cup minced ham or bacon (optional)
Salt and freshly ground pepper to taste
6 tablespoons olive or vegetable oil

Combine rice, spinach, eggs, cheese, optional ham or bacon and salt and pepper. In skillet heat oil and drop in heaping tablespoons of the rice-spinach mixture. Fry 2 to 3 minutes on each side or until golden. Serve hot.

Serves 4

Green Rice with Shrimp Newberg

GREEN RICE

4 cups cooked rice
¾ cup chopped parsley
2 tablespoons chopped chives
½ cup melted butter
3 eggs, separated

Combine rice, parsley, chives, butter and egg yolks. Beat egg whites until stiff and fold into rice mixture. Pour into buttered ring mold and bake in preheated 325° oven 25 minutes. Unmold on warm platter and fill center with shrimp Newberg.

Cook the shrimp in boiling water 3 minutes. Drain immediately, cool and shell. Heat butter in saucepan and add flour. Cook and stir 3 minutes. Do not burn. Gradually add milk, stirring constantly, and cook until smooth and slightly thickened. Add mustard, cayenne and Madeira. Place the saucepan over boiling water and continue cooking, stirring occasionally, 15 minutes. Beat the egg yolks with the cream, beat in ½ cup of hot sauce and gradually return to rest of sauce, stirring constantly, add reserved shrimp and cook until heated through. Pour into green rice ring and serve immediately.

Serves 6

Note: Any other creamed dish such as chicken, turkey, sweetbreads, tuna, salmon, oysters or sweetbreads may be substituted for the shrimp Newberg.

Hopping John

A traditional New Year's Day dish eaten throughout the South, Hopping John carries a special omen of good luck.

2 cups black-eye peas
1 ham bone, cracked or
½ pound salt pork, blanched and sliced
1 bay leaf
1 onion, chopped
4 cups hot cooked rice

Soak peas in water to cover 4 hours. Drain and in saucepan combine with 2 quarts water, ham bone or salt pork, bay leaf and onion. Bring to boil, skim off any scum that rises to surface, reduce heat and simmer 1½ hours or until peas are tender. Serve rice separately or combined with peas as an accompaniment to meat, poultry or collard greens.

Serves 8

SHRIMP NEWBERG

1½ pounds raw shrimp
3 tablespoons butter
2 tablespoons flour
2 cups milk
½ teaspoon dry mustard
Pinch cayenne pepper
¼ cup Madeira
2 egg yolks
1 cup heavy cream
Salt and freshly ground pepper to taste

Jambalaya

Like the Spanish paella, its ancestor, jambalaya is simply rice and whatever meat or seafood combination that suits one's taste. No two Creole cooks will prepare identical dishes.

2 cups long grain rice, washed and drained
1 onion, chopped
2 garlic cloves, minced
3 tablespoons vegetable oil
1 green pepper, diced
3 cups peeled and chopped tomatoes
¼ teaspoon cayenne pepper
¼ teaspoon thyme
¼ teaspoon powdered cloves
1 pound raw ham, cut in ½-inch dice
Salt and pepper
1 pound raw shrimp, shelled and deveined
Chopped parsley

In large skillet sauté onions and garlic in oil 5 minutes. Add green pepper, tomatoes, cayenne, thyme, cloves and 2½ cups water. Bring to rapid boil and add ham and rice. Cover, lower heat to simmer and cook 20 minutes. Season to taste with salt and pepper. Place shrimp on top, cover and continue cooking 10 minutes longer. Garnish with chopped parsley.

Serves 6 to 8

Red Beans and Rice

A classic Creole dish which appears on the menu of many New Orleans restaurants on Mondays only. A good way to use up that ole ham bone!

1 pound dried small red beans, washed, soaked in water to cover overnight and drained
1 ham bone, cracked to release marrow during cooking
1 onion, chopped
3 garlic cloves, minced
Salt and freshly ground pepper to taste
6 cups freshly cooked long grain rice
1 cup chopped green onions

Drain beans and set aside. Place ham bone in large pot with 2 quarts water. Bring to boil and skim off any scum that rises to surface. Add beans, onion and garlic. Cover, bring to boil and simmer 2 hours or until beans are soft, adding boiling water as needed. When beans are done, remove ham bone and mash some of the beans with the back of a spoon against the side of the pot to thicken the remaining liquid. Remove meat from bone and return to pot. Add salt and pepper to taste. Spoon beans onto individual plates and let everyone add his own mound of rice and green onions. Serve with green salad with tart vinaigrette dressing.

Serves 6

Chou Farci
(Stuffed Whole Cabbage)

The French have a delightful and delicious way of serving stuffed cabbage dishes. This is one of my favorites — simple to do and beautiful to serve.

1 large Savoy (curly) cabbage
1 cup long grain rice, washed and drained
1 pound ground veal
1 pound ground pork butt
1 cup finely minced ham
2 garlic cloves, minced
½ teaspoon thyme
1 teaspoon salt
½ teaspoon freshly ground pepper
2 onions, thinly sliced
2 tablespoons butter
4 carrots, scraped and sliced
3 turnips, scraped and sliced
1 bay leaf
3 cups beef or veal stock
1 cup dry white wine
1 cup fresh peas

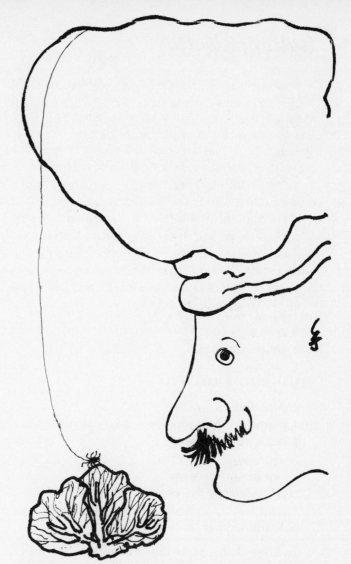

Remove any tough outer leaves from cabbage and plunge into boiling water for 5 minutes to limp leaves. Drain and cool. Combine rice, meats, garlic, thyme, salt and pepper and mix well. Beginning at center of cabbage and finishing toward outer leaves, stuff meat mixture into leaves. Leave the outside leaves unstuffed and form cabbage into its original shape. Tie firmly with string and set aside. Sauté the onions in the butter 5 minutes. Add carrots, turnips, bay leaf, stock and wine. Bring to boil. Place the prepared cabbage in a large soup pot or canning pot and pour vegetables and stock over. Cover and simmer 2 hours or until cabbage center is tender when pierced with a fork. Add the fresh peas the last 10 minutes of cooking. To serve, transfer cabbage to heated platter, remove string and surround with vegetables from the pot. Cut in wedges and serve with the vegetables. The broth may be served in individual soup cups along with the cabbage. Fresh fruit and cheese are the perfect ending for this meal.

Serves 6 to 8

Italian Risotto

Rice cuisine in Italy comes mainly from the areas of Venice and Milan. An Italian risotto is far more moist than a Spanish paella or a Creole jambalaya. There is less chance of the cooked rice turning into a mush if Italian rice, such as Arborio* or Vialone*, is used. The secret to making a good risotto is to add hot liquid, a cup at a time, while cooking, letting it evaporate before adding more liquid. The process is repeated, cooking over medium heat, until rice is just tender and still retains its shape and the sauce is thick and creamy. In Venice, risotto is always served with a meat or seafood dish. In Milan, the risotto is often the main meal. There are endless combinations: like paella and jambalaya, every risotto can be as different as its cook.

Available in Italian food stores. Converted rice or brown rice may be substituted.

MUSHROOM RISOTTO

2 cups Italian rice
1 pound fresh mushrooms, thinly sliced
1 onion, thinly sliced
½ cup butter
1 cup dry white wine
Approximately 5½ cups hot chicken stock
½ cup freshly grated Parmesan cheese
Salt and pepper to taste

In large skillet, sauté mushrooms and onion in half the butter until limp. Add rice, stirring until rice takes on color of the mushrooms and onion. Pour wine over and cook over medium heat until most of liquid has evaporated. Add 1 cup of the stock, simmer until evaporated, and continue process until rice is tender. Add remaining butter, cheese and salt and pepper. Let rest for 2 to 3 minutes and serve with additional Parmesan cheese.

Serves 6

CHICKEN LIVER AND VEAL RISOTTO

2 cups Italian or long grain rice
4 tablespoons each butter and olive oil
1 onion, thinly sliced
1 garlic clove, minced
½ pound chicken livers, halved
½ pound ground veal
½ cup dry white wine
1 carrot, finely chopped
1 stalk celery, finely chopped
Approximately 6 cups hot chicken stock
Salt and pepper to taste
Freshly grated Parmesan cheese

In skillet sauté onion and garlic in half the butter and oil until transparent. Add chicken livers and veal and sauté until livers are just cooked through. Add wine, turn off heat, cover and set aside. In another skillet, heat remaining butter and oil and sauté the rice until grains are well coated with oil. Add carrot and celery and 1 cup of the stock, stir and cook over low heat until liquid has evaporated. Continue this process (see Italian risotto) until rice is tender. Combine with the meat mixture, season with salt and pepper and heat through. Let rest 2 to 3 minutes and serve with Parmesan cheese.

Serves 6

RISOTTO MILANESE (RICE WITH SAFFRON)

2 cups Italian or long grain rice
1 onion, chopped
3 tablespoons bone marrow, minced
½ cup butter
½ cup dry white wine
6 cups very hot chicken stock
⅛ teaspoon saffron
Salt and pepper to taste
Freshly grated Parmesan cheese

In skillet sauté onion in marrow and half the butter until transparent. Add rice and sauté until rice is golden and well coated with butter. Add wine and simmer until evaporated. Add 1 cup of hot stock and cook until evaporated. Steep saffron in another cup of stock, add to rice, stir well to blend, and cook until liquid has evaporated. Continue this process (see Italian Risotto) until rice is tender. Add remaining butter and salt and pepper. Serve with Parmesan cheese. In Italy, Risotto Milanese is often accompanied by Osso Buco or other braised meat dishes.

Serves 6

FRIED RISOTTO

This an excellent way to use up any remaining rice, whether it be risotto, pilaf or plain.

In heavy skillet over medium heat, melt some butter and/or oil. Add rice, flatten with spatula and cook until a golden crust begins to form on bottom. Be sure crust does not burn. When crisp, carefully turn upside down onto plate, add more butter or oil to pan and slide rice crust back into pan to crisp the underside. Cut in wedges and serve with freshly grated Parmesan cheese as main dish accompaniment or for brunch with broiled tomatoes and sausages.

Risi E Bisi

A specialty of Venice when fresh peas are first harvested and considered by the Italians as a very thick soup, Risi e Bisi is eaten with a fork, not a spoon. Serve as an accompaniment to sautéed sausages, chicken livers, veal or shellfish.

2½ cups rice
½ cup diced lean bacon
3 tablespoons chopped green onions
3 tablespoons each butter and olive oil
6 to 8 cups hot chicken stock
3 cups fresh peas
Salt to taste
Freshly grated Parmesan cheese
3 sprigs parsley, chopped

Sauté the bacon and onions in butter and oil until bacon begins to brown. Add the rice, and stirring constantly, cook until golden. Lower heat to simmer and gradually add 1 cup of hot stock. When absorbed, add another cup of the hot stock and the peas. Continue in this manner until the rice is tender and soup is of desired consistency. Add salt to taste. Serve hot with Parmesan cheese and chopped parsley.

Serves 6 to 8

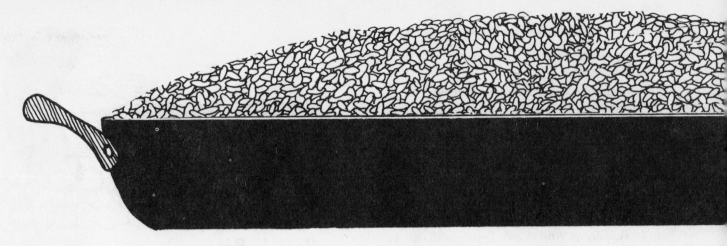

Paella

Named for the pan "paella" in which this famous rice dish is cooked, there are hundreds of variations. The three essentials are rice, olive oil and saffron. Paella may be simple or elaborate, depending upon the seafood or meats included.

3 cups long grain rice
1 chicken fryer, cut up
½ cup olive oil
½ pound chorizo sausage, blanched and cut into ½-inch slices
1 small onion, chopped
2 garlic cloves, minced
1 sweet red or green pepper, cut in thin strips
1 cup peeled, seeded and chopped tomatoes
¼ teaspoon saffron dissolved in ¼ cup hot water
6 cups boiling water
½ pound raw shrimp, shelled and deveined, tails intact
6 or more clams in shells, well scrubbed
6 or more mussels in shells, well scrubbed
½ cup fresh peas
Salt and freshly ground pepper to taste
Lemon wedges

Sprinkle chicken pieces with salt and pepper and brown on all sides in half the olive oil. Remove and set aside. In same oil, brown chorizo slices and remove. Add remaining oil to skillet and fry onions, garlic and pepper 5 minutes. Add tomatoes and cook until liquid has evaporated. Add rice, stirring well to coat with oil, and cook until lightly golden. Blend in saffron water and pour 2 cups of the boiling water over. Cook over high heat until rice has absorbed the liquid. Do not stir. Pour 2 more cups boiling water over rice and cook until water is absorbed. Remove from heat, pour last 2 cups of boiling water over rice and place chicken pieces, shrimp, clams, mussels and peas on top of rice, finish cooking in a preheated 400° oven 20 minutes. Serve immediately from the pan with garnish of lemon wedges.

Serves 6 to 8

Any of the following may be added to a paella:

- artichoke hearts, asparagus tips, green beans.
- lobster, squid, eel, firm fish fillets.
- ham, bacon, pork.
- rabbit, giblets, sweetbread, veal kidneys, other fowl.

Adobo

A popular chicken and pork dish from the Philippines

1 chicken fryer, cut up
1½ pounds lean pork butt, cut into 1-inch cubes
½ cup each vinegar and water
2 bay leaves
1 teaspoon salt
½ teaspoon freshly ground pepper
3 tablespoons lard
2 onions, sliced
2 garlic cloves, minced
Water or sherry for deglazing
Freshly cooked rice

In stew pot combine chicken, pork, vinegar, water, bay leaves, salt and pepper. Bring to boil, cover and simmer 45 minutes or until meats are tender. Uncover and cook until liquid is evaporated. Add lard and brown chicken pieces, pork, onions and garlic. Deglaze pot with a little water or sherry. Heat through and serve over freshly cooked rice.

Serves 6 to 8

Picadillo (Cuban Stew)

1½ pounds beef chuck, cut in ½-inch cubes
1½ pounds pork butt, cut in ½-inch cubes
6 tablespoons lard
3 garlic cloves, minced
3 onions, chopped
1 green pepper, seeded and chopped
2 sweet red peppers, seeded and chopped
2 bay leaves, crumbled
1 teaspoon dried oregano leaves
1 teaspoon cumin seed
6 tomatoes, peeled and chopped
½ cup lemon juice
1 cup raisins
1 cup blanched almonds
2 teaspoons salt
1 teaspoon freshly ground pepper
1 cup dry red wine
1 cup beef stock

In large skillet brown the beef and pork in 2 tablespoons of the lard. Remove to oven-proof casserole. Sauté the garlic and onions in remaining lard until onions become transparent. Add the peppers, bay leaves, oregano, cumin and tomatoes. Bring to rapid boil and cook 5 minutes. Add the lemon juice, raisins, almonds, salt and pepper and pour over the meat, stirring well to blend. Pour the wine and stock over all and place in a preheated 300° oven for 2 hours or until meat is tender. Serve with Cuban Rice (page 22).

Serves 8

Rice Guisado

From Puerto Rico comes this mixed seafood and meat stew with rice.

1 cup rice, soaked in water to cover 20 minutes and drained
2 onions, chopped
2 garlic cloves, minced
1 tablespoon lard
1 pound firm fish fillets, cut in 1-inch cubes
½ pound lobster meat, cut in chunks, or
½ pound scallops, halved if large
½ pound ham, cut in ½-inch dice
¼ pound chorizo sausage, sliced ½ inch thick
1 bay leaf
1 or more chili peppers
2 quarts chicken stock or water
2 tablespoons chopped coriander (cilantro) or parsley
Lemon or lime wedges

Sauté onions and garlic in lard for 2 minutes. Add fish, lobster or scallops, ham and chorizo. Continue sautéing 5 minutes. Add rice, bay leaf, chili pepper and stock. Bring to boil, lower heat to medium and cook 15 minutes or until rice is tender. Serve in large stew bowls and garnish with coriander. Pass lemon or lime wedges.

Serves 6

Tortilla de Arroz (Rice Omelet)

1 cup cooked rice
6 eggs, separated
¼ cup milk
Pinch of powdered saffron
Salt and freshly ground pepper
2 tablespoons butter
2 cups grated Monterey Jack cheese
3 tablespoons or more diced green, canned chilis (optional)

Combine egg yolks, milk and saffron and beat well. Add rice and salt and pepper. Fold in stiffly beaten egg whites. Heat butter in an omelet pan, pour in egg-rice mixture and cook over medium heat until puffy. Sprinkle with grated cheese and optional chilis and place under broiler to melt cheese and brown lightly. Cut into wedges and serve immediately. Great for breakfast or brunch with a rasher of bacon and a broiled tomato.

Serves 4 to 6

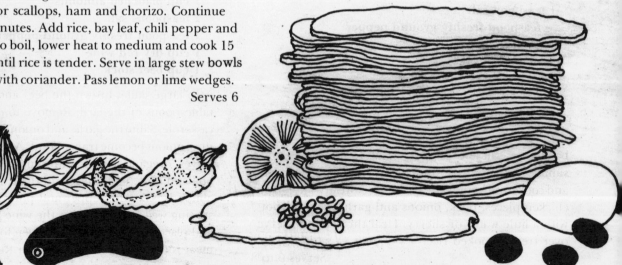

Arroz con Pollo

This chicken and rice dish and its many variations are very popular in Puerto Rico and Mexico.

1½ cups long grain rice
1 3- to 4-pound chicken, cut up
4 tablespoons lard or
¼ cup olive oil
2 garlic cloves, minced
1 bay leaf, crumbled
2 tablespoons chopped parsley or fresh coriander
1 teaspoon salt
½ teaspoon freshly ground pepper
½ cup pitted black olives
¼ cup capers
½ cup diced fresh sweet pepper or pimiento

In large skillet or Dutch oven, brown chicken pieces on all sides in lard or oil. Add garlic, bay leaf, parsley or coriander, stirring well with chicken, and the rice. Blend in 3 cups hot water, salt and pepper. Bring to boil, lower heat to simmer and cover. Cook undisturbed for 30 minutes. Remove from heat, toss in olives, capers and peppers and cover, and let rest for 10 minutes.

Serves 4 to 6

Variation: Arroz con Pollo Mejicano. Omit water and add 1½ cups peeled, chopped tomatoes, 1½ cups chicken stock, 1 teaspoon mild California chili powder and ½ teaspoon oregano.

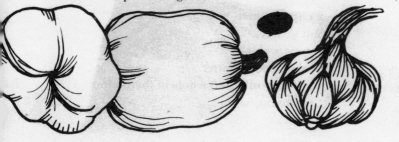

Pastel de Lujo

A "rice meat cake" from the Yucatan. Similar to a tamale but prepared in a casserole.

MEAT FILLING

1 pound pork, cut into ½-inch dice
1 pound raw chicken meat, cut into ½-inch dice
1 onion, chopped
1 or more green chili peppers, seeded and chopped
3 tablespoons lard or oil
½ cup pitted black olives
1 tablespoon capers
2 tablespoons raisins
1 cup tomato puree
½ teaspoon cinnamon
¼ teaspoon each allspice and ground cloves
½ cup Madeira
Salt and freshly ground pepper to taste

RICE-CORN MIXTURE

4 cups cooked rice
3 cups fresh corn kernels
¾ cup milk
2 eggs, beaten
4 tablespoons butter, melted
1 teaspoon salt

Brown the meats, onion and chili in the lard. Add the remaining meat filling ingredients and cook 15 minutes or until most of the liquid has evaporated. Set aside. Puree the corn and milk in blender until smooth. Combine with the rice, eggs, butter and salt. In a large buttered casserole, put half the rice-corn mixture, add the meat mixture and top with remaining rice-corn mixture. Bake in a preheated 350° oven 45 minutes.

Serves 8

Kitcherie

Originally an Indian breakfast food of rice and lentils with various curry spices. The British Colonials substituted fish for the lentils and renamed it Kedgeree.

1 cup rice
1 cup lentils, mung beans or yellow peas, soaked and
   drained
1 onion, thinly sliced
2 garlic cloves, minced
6 tablespoons clarified butter
¼ teaspoon each powdered cloves, nutmeg and
   ginger
3 cardamon seeds, crushed
½ teaspoon freshly ground pepper
1 teaspoon salt
3 hard-cooked eggs, quartered

Sauté onion and garlic in clarified butter 5 minutes. Add spices and salt and continue sautéing another 3 minutes. Add rice and lentils, stirring well to coat well with butter. Cook and stir 3 minutes and add 2½ cups boiling water. Cover and simmer 20 minutes or until rice and lentils are tender. If necessary, add a little more boiling water. Mixture should be dry but thoroughly cooked. Serve garnished with quartered eggs.

Serves 6

Kedgeree: Omit lentils and reduce water to a total of 2 cups. Just before serving add 2 cups cooked flaked fish to rice and heat through.

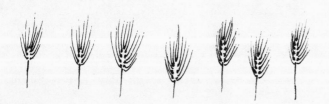

Mujadarra (Lentils and Rice)

A medieval Persian dish of the poor, Mujadarra is good, wholesome and humble food.

1 cup long grain rice, washed and drained
1 cup brown lentils, washed and drained
3 onions
2 tablespoons each butter and oil
Salt and freshly ground pepper
Yoghurt

Boil lentils in 1 quart water 40 minutes or until tender. Chop 1 of the onions and sauté in 1 tablespoon each of the butter and oil. Combine with lentils, rice, salt and pepper, mixing well and adding additional water if needed. Lower heat to simmer, cover and continue cooking 20 minutes or until rice is tender. Slice the remaining 2 onions thinly and fry until golden in the remaining butter and oil. Serve the rice and lentils garnished with the fried onions. Pass a bowl of yoghurt. Complete the meal with cool, sweet melon for dessert.

Serves 6

Chilau (Persian Steamed Rice)

Chilau Kebab, steamed rice with broiled meat, is the national dish of Iran. The basmati rice is painstakingly prepared, sometimes a day in advance, and is the base or accompaniment for almost every dish in Iran.

2 cups basmati rice*
Salt
7 cups water
8 tablespoons butter
4 to 6 egg yolks, left whole in shells after
   removing whites

Pick out and discard any dark grains in the rice. Wash in warm water until water runs clear. Cover with lightly salted water and let stand 6 hours or overnight. An hour before serving, drain rice and rinse in cold water. In at least a 3-quart saucepan, bring 7 cups water and 1 teaspoon salt to a rolling boil. Gradually add the rice so the water remains boiling. Continue boiling for 7 minutes or until grains are tender between fingers when tested. Drain and rinse with warm water. Drain well. Melt 4 tablespoons of the butter in heavy pan and add 2 tablespoons hot water and the rice. Place a cloth napkin over the pot, making sure the napkin corners are folded over to prevent burning, and cover with lid. Simmer on lowest heat for 25 minutes. The rice should be fluffy and the crust on the bottom crisp and golden. Toss in remaining butter. Mound rice on a plate, surround with broken crust and place kebabs on top. Or serve as a rice dish to accompany any other meat. An egg yolk is often served with each serving of rice to be mixed with the rice and a lump of butter. The result is almost a creamed rice. Delicious.

Serves 4 to 6

*Basmati rice is available in Middle Eastern food stores. Long grain rice may be substituted. Reduce soaking time to 1 hour.

Persian Chicken Kebab

4 chicken breasts, halved
4 chicken legs and thighs, split
½ cup lemon juice
1 tablespoon grated lemon peel
1 onion, finely minced
1 teaspoon salt
¼ cup melted butter
⅛ teaspoon saffron threads dissolved in
1 tablespoon hot water

Marinate the chicken pieces in the lemon juice, lemon peel, onion and salt at room temperature 2 hours. Put chicken pieces on long skewers, brush with melted butter and saffron water and broil over hot coals, basting and turning frequently. Meat is done when pierced with a fork and clear, not pink, juices run. Serve with Chilau (page 62).

Serves 4 to 6

Dolmas Tomates (Stuffed Tomatoes)

Rice is often used as the base for stuffing Middle Eastern dishes. This combination of green vegetables and rice in a tomato shell is most pleasing to the eye and palate.

8 large ripe tomatoes
½ cup long grain rice, washed and drained
3 tablespoons olive oil
½ cup chopped onions
1 garlic clove, minced
¼ teaspoon allspice
¼ teaspoon freshly ground pepper
½ teaspoon salt
1 cup chopped zucchini
1 cup chopped spinach
2 tablespoons chopped fresh mint or
2 tablespoons dried mint
2 tablespoons currants

Slice tops off tomatoes, carefully scoop out pulp, turn upside down and drain until ready to fill. Heat the oil in large skillet and sauté onions and garlic 5 minutes. Add rice, allspice, pepper and salt. Continue sautéing until rice is a light golden color. Pour in 1 cup hot water and cook over medium heat, uncovered, until water has evaporated. Add zucchini, spinach, mint and currants and mix well. Cover with tight-fitting lid, reduce heat to lowest simmer and cook 15 minutes. Fill prepared tomato shells with rice mixture and place in shallow baking dish. Bake in preheated 325° oven 20 minutes. Serve with shish kebab or roasted leg of lamb.

Serves 8

Eggplant, Tomatoes and Rice Algerian

2 cups hot cooked rice (preferably cooked in stock)
1 eggplant, sliced ½-inch thick
3 ripe tomatoes, sliced ½-inch thick
Flour seasoned with
Salt, pepper, oregano and sweet basil
Vegetable oil for sautéing
Mushroom sauce
Watercress sprigs for garnish

Have the hot rice ready on a warm platter. Dredge the eggplant and tomato slices in seasoned flour and sauté in hot oil until golden. Place alternate slices of eggplant and tomatoes on top of rice and pour the creamed mushroom sauce over all. Garnish with watercress sprigs.

Serves 4

Mushroom Sauce

- **1 cup sliced fresh mushrooms**
- **2 tablespoons butter**
- **2 shallots or ½ onion, chopped**
- **1 tablespoon flour**
- **¾ cup rich milk**

Sauté the mushrooms in butter 3 minutes, add shallots or onion and flour. Mix well. Blend in milk. Continue cooking and stirring until sauce is thickened.

Turkish Pilaf

- **2 cups long grain, brown or white rice**
- **1 onion, finely chopped**
- **½ cup butter or half butter and oil**
- **½ teaspoon cinnamon**
- **¼ teaspoon each allspice and mace**
- **Pinch of powdered cloves**
- **4½ cups beef or chicken stock**
- **Salt and freshly ground pepper to taste**
- **½ cup pinenuts, pistachio nuts, or slivered almonds in any combination**
- **¼ cup currants**

Sauté onion in half the butter and/or oil until transparent. Add rice, stirring constantly, and sauté 5 minutes. Add spices and stock. Bring to rapid boil, lower heat to simmer and cover. Cook 15 minutes for white rice, 35 minutes for brown, or until tender. Salt

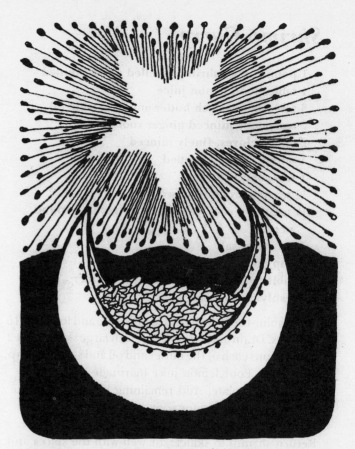

and pepper to taste. In another pan, sauté nuts and currants in remaining butter and/or oil. Mix with rice and serve with lamb, chicken or other main dish.

Serves 6

Variations: **Simple Pilaf.** Omit spices and nuts.

Herb Pilaf. Omit spices and add:

- ½ teaspoon each dried thyme and marjoram.
- 1 teaspoon chopped chives and 1 tablespoon chopped parsley or fresh coriander.
- ½ teaspoon oregano and a pinch of sage.

Shrimp Curry

1 pound large shrimp, shelled and deveined
1 tablespoon lemon juice
3 tablespoons each butter and vegetable oil
2 teaspoons minced ginger root
2 garlic cloves, finely minced
1 onion, finely chopped
1 teaspoon turmeric
½ teaspoon cumin powder
¼ teaspoon each cayenne and freshly ground
  pepper
1 teaspoon salt
1 cup coconut milk (page 27)
3 tablespoons chopped fresh coriander
Freshly cooked rice

Combine the shrimp and lemon juice and let stand 15
minutes. Drain and reserve juices. In large skillet heat 2
tablespoons each of the butter and oil and sauté shrimp
2 minutes. Pour lemon juice marinade over and
transfer to a plate. Add remaining butter and oil to
skillet and sauté ginger root, garlic and onions 5
minutes. Add remaining spices and cook 1 minute.
Return shrimp to skillet, stir well with the spices and
seasonings and add coconut milk and half the chopped
coriander. Cover and cook 5 minutes. Do not overcook
shrimp. Serve immediately over freshly cooked rice and
sprinkle with remaining chopped coriander.
Accompany with chutney.

Serves 4

Masala Dar Pilau (Spiced Rice)

Pilau, pellao, pulao, pilav, pilaf and pilaff are familiar
throughout India and the Middle East. They may be
simple or elaborate with the addition of meat, poultry,
vegetables, eggs, dried beans or lentils. As with other
famous rice dishes of the world, the final outcome
depends on the availability of ingredients and the
preferences of the cook.

2 cups long grain rice
2 onions, thinly sliced
2 garlic cloves, finely minced
6 tablespoons clarified butter or half butter and half
  vegetable oil
1 cup yoghurt
4 cardamon seeds, crushed
½ teaspoon ground turmeric
½ teaspoon ground ginger
⅛ teaspoon cayenne pepper
1 teaspoon salt
3 cups hot chicken stock
Green pepper slices
Tomato wedges
Coriander sprigs

Soak rice in water to cover 30 minutes and drain.
Sauté onion and garlic in butter and/or oil until
transparent, stir in rice and continue cooking and
stirring 5 minutes to coat rice well with butter. Blend in
yoghurt, cardamon seeds, turmeric, ginger, cayenne
and salt. Add broth and bring to rapid boil without
stirring. Lower heat to simmer, cover and cook 20
minutes or until rice is tender. Transfer to platter and
garnish with green pepper, tomato and coriander.
Serve as accompaniment to roast chicken, shish kebabs,
broiled fish or other main dishes.

Serves 8

Shahjahani Biryani (Saffron Rice with Spiced Lamb)

2 cups long grain rice, washed and drained
¼ teaspoon saffron threads dissolved in
¼ cup hot water
½ cup hot water
½ cup clarified butter
2 onions, thinly sliced
¼ cup unsalted cashews
¼ cup blanched, slivered almonds
⅓ cup raisins
2 teaspoons minced ginger root
2 garlic cloves, minced
½ teaspoon cumin seed
2 pounds boneless lamb, cut in 1-inch cubes
1 stock cinnamon
4 whole cloves
¼ teaspoon each cayenne pepper, cardamon seeds
   and freshly grated nutmeg
1½ cups chicken stock

1 cup plain yoghurt
Salt and freshly ground pepper

Bring 4 cups water to boil and, stirring constantly, gradually add the rice. Cook for 10 minutes, drain and place in bowl. Combine with saffron thread water and set aside. In large Dutch oven, heat ¼ cup of the clarified butter and fry the onions until golden. Remove onions with slotted spoon and reserve. In same butter, fry the nuts approximately 1 minute. Remove with slotted spoon and set aside. Repeat with raisins. Add 2 more tablespoons butter to Dutch oven and cook the ginger root, garlic and cumin 1 minute. Add and brown the meat on all sides. Blend in remaining spices, ¾ cup of the stock and the yoghurt. Cover and simmer 20 minutes. Carefully spoon rice evenly over the lamb mixture, sprinkle remaining stock over rice and drizzle remaining butter on top. Cover and continue cooking on low heat for 15 minutes. To serve, mound the rice and lamb on a large platter, remove cinnamon stick and sprinkle top with reserved onions, nuts and raisins.

Serves 8

Baked Whole Pumpkin With Rice And Lamb

The pumpkin shell becomes the edible "casserole", imparting a delicious, subtle flavoring to this stew.

- 1 cup cooked long grain brown rice
- 1 medium-sized pumpkin
- 3 tablespoons butter
- 1 pound boneless lamb, preferably from the leg, cubed
- 1 onion, chopped
- 1 celery rib, including tops, chopped
- 1 apple, peeled, cored and diced
- ¼ cup raisins
- 1 teaspoon grated orange peel
- ½ teaspoon cinnamon
- ¼ teaspoon each nutmeg and cloves.
- 2 cups apple cider
- ¼ cup toasted slivered almonds
- Salt & freshly ground pepper to taste

Cut 1 inch off the top of the pumpkin and reserve for lid. Remove seeds and carefully scoop out 1 cup of pumpkin pulp. Dice and reserve. In heavy skillet, melt butter and brown the meat and onions 10 minutes. Add celery, reserved pumpkin pulp, apple, raisins, orange peel and cinnamon. Blend well with meat mixture and cook 2 minutes. Pour cider over all, bring just to boil, lower heat to simmer, cover and cook 20 minutes. Remove from heat, add rice and pour into pumpkin shell. Cover with reserved lid, place pumpkin on oiled shallow baking dish and bake in preheated 350° oven 2 hours or until pumpkin is tender. Just before serving, sprinkle the almonds over the stew.

Serves 4 to 6

Mixed Fruit Pilaf

Rice pilaf crowned with a colorful fruit sauce to be served with roast fowl or lamb, or barbecued meats.

1 cup long grain rice
6 tablespoons butter
2 cups hot water
Salt to taste
1/3 cup dried apricots, quartered
1/3 cup dried pitted prunes, quartered
¼ cup currants
¼ cup slivered almonds
2 tablespoons each honey and hot water, mixed well

Melt 3 tablespoons of the butter in heavy skillet and stir in rice to coat thoroughly; do not brown. Add water, bring just to boil, lower heat to simmer, cover with tight-fitting lid and cook 20 minutes or until rice is tender. Season with salt to taste. In another skillet, melt remaining butter and add apricots, prunes, currants and almonds. Saute, stirring constantly, until lightly browned. Blend in honey-water mixture and cook over low heat 7 minutes or until thickened. Spoon hot rice onto warmed serving dish and top with fruit sauce.

Serves 4

Yoghurt and Mint Rice

1 cup brown rice
3 tablespoons butter
1 onion, minced
2½ cups hot chicken stock
¼ cup chopped fresh mint
3 tablespoons chopped parsley
3/4 cup yoghurt
Salt and freshly ground pepper to taste

Heat butter and saute onion until transparent. Stirring constantly, add rice and saute 5 minutes. Blend in stock, bring to rapid boil, cover, lower heat to simmer and cook 30 minutes or until rice is tender. Stir in mint, parsley, yoghurt, salt and pepper. Cover with tea towel and let stand 5 minutes.

Serve 4

Danish Liver and Rice Pudding

A typical Danish family dish, serve with apple coleslaw and rye bread.

1 cup rice
4 cups milk
1 teaspoon salt
1 onion, chopped
3 tablespoons butter
2 cups chopped cooked beef liver
2 eggs, beaten
3 tablespoons molasses
¼ teaspoon each ground allspice, cloves and
  black pepper
½ cup raisins
Melted butter

Combine rice, milk and salt in top of double boiler. Cover and cook over boiling water for 1 hour or until rice is tender and mixture is thick. Let cool. Saute onion in butter for 3 minutes. Remove from heat and blend in liver, eggs, molasses, spices, raisins and rice mixture. Pour into buttered 2-quart casserole. Cover and bake in a preheated 325° oven 1 hour. Serve with melted butter.

Serves 6

Vegetarian Rice and Nut Roast

2 cups cooked brown rice
1 cup chopped mixed nuts
1 cup whole wheat bread crumbs
½ cup milk
1 egg, beaten
1 onion, chopped
6 tablespoons butter, melted
¼ cup chopped parsley
1 teaspoon chopped fresh sage (or ¼ teaspoon dried)
1 teaspoon chopped fresh thyme (or ¼ teaspoon dried)
1 teaspoon Worcestershire sauce
Salt and freshly ground pepper to taste

Combine ingredients well and shape into loaf. Place in buttered pan and bake in preheated 375° oven 35 minutes. Serve with Cumberland Sauce (following).

Serves 4

Cumberland sauce:

2 tablespoons butter
3 shallots, minced
1 orange
1 lemon
½ cup red currant jelly
¼ cup port
Pinch powdered ginger
Pinch cayenne pepper

In heavy skillet, melt butter over low heat. Saute shallots gently 3 minutes; do not brown. Add the juice of the orange and lemon, the peels, cut into thin julienne, and remaining ingredients. Stir well, bring just to boil and remove from heat.

Makes 1 cup

Rice Frittata

Served warm or cold, this rice cake is a great Italian favorite. Excellent picnic fare!

2 cups cooked rice
3 cups cooked, well drained chopped vegetables such as spinach, zucchini and/or asparagus
2 tablespoons chopped green onions
½ teaspoon crushed oregano leaves
¼ teaspoon crushed rosemary leaves
Pinch sage powder
½ cup grated Parmesan cheese
3/4 cup grated Monterey Jack cheese
4 eggs, beaten
¼ cup olive oil
Salt and freshly ground pepper to taste

Combine ingredients and pour into a buttered 2-quart baking dish or pan. Bake in a preheated 350° oven 25 minutes or until eggs have set.

Serves 6

Variation: Add ½ cup minced cooked chicken, ham or veal to mixture before baking.

Armenian Stuffed Melon

A typical Middle Eastern rice, lamb and nut stuffing baked in a whole Persian melon. Beautiful to serve, delicate and delicious in flavor.

½ cup long grain rice
1 Persian melon or large cantaloupe
3 tablespoons butter
1 onion, chopped
½ pound ground lamb or raw chicken meat
¼ cup pine nuts
¼ cup currants
¼ cup honey
1 cup hot water
¼ teaspoon cinnamon
2 tablespoons sugar
Salt to taste

Cut 1 inch off top of melon and reserve for lid. Discard seeds and scoop out 1 cup of melon pulp. Chop and reserve. Heat butter in skillet and saute onion until transparent. Add meat and brown lightly. Blend in rice, nuts, currants and melon pulp. Combine honey, water and cinnamon. Stir into meat mixture and cook over medium heat until liquid is absorbed. Season with salt, remove from heat and cool to lukewarm. Sprinkle inside of melon with sugar and stuff with meat-rice mixture. Replace lid and secure with toothpicks. Place on oiled ovenproof dish and bake in preheated 350° oven 1 hour or until melon is tender.

Serves 4

Yugoslavian Style Stuffed Peppers

1 cup cooked rice
8 green peppers
1 onion, chopped
4 tablespoons olive oil
½ pound ground veal
½ pound ground lean pork
¼ cup chopped celery
2 tablespoons minced parsley
1 cup chopped peeled tomatoes
2 teaspoons paprika
1 teaspoon sugar
Salt and freshly ground pepper to taste
1 cup tomato juice

Cut off stem end of peppers and remove core and seeds. Reserve tops. Saute onions in 2 tablespoons of the olive oil 2 minutes. Add veal, pork, celery and parsley. Cook and stir 10 minutes or until meats lose their red color. Add tomatoes and bring just to boil. Blend in rice, paprika, sugar, salt and pepper. Remove from heat. Arrange peppers in heatproof dish, fill with meat-rice mixture to within 1 inch of top and replace reserved lids. Drizzle remaining olive oil over all and pour in tomato juice. Cover with aluminum foil and bake in preheated 325° oven 1 hour. Serve hot or cold.

Serves 4

Haw Yip Faahn (Steamed Lotus Leaf Rice)

The lotus leaves impart a richly fragrant aroma to the rice and filling.

3 cups long grain rice (or half glutinous rice)
¼ cup peanut oil
½ pound Chinese barbecued pork, diced
3 lop chiang*(Chinese sausages), rinsed and diced
¼ pound lop yook*(Chinese smoked bacon) or bacon, diced
6 dried scallops*, soaked in hot water to soften and shredded with fingers
½ cup diced water chestnuts

1 cup diced bamboo shoots
6 dried forest mushrooms, soaked to soften and diced
2 tablespoons soy sauce
½ teaspoon Oriental sesame oil
2 green onions, chopped
Salt to taste
Fresh coriander sprigs
12 dried lotus leaves**, soaked to soften, washed well and drained

Cook rice as for Chinese steamed rice (page 20). Keep hot. In large skillet or wok heat peanut oil and stir-fry meats, scallops, water chestnuts, bamboo shoots and mushrooms for 5 minutes. Add soy, sesame oil and HOT rice and combine well, turning rice and other

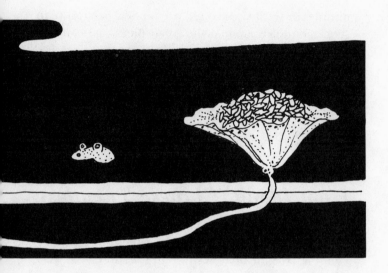

ingredients all the while for 3 minutes. Add the green onions and salt. Remove from heat and cool to lukewarm. Place about 1 cup of filling on each lotus leaf. Lay a coriander sprig on top, turn up stem edge to cover filling, bring sides over and fold remaining edge to completely envelop rice filling. Tie in place with string if package has not been wrapped neatly enough. Stack seam side down in a steamer basket or heatproof plate. Repeat until all filling and leaves are used. Steam above water for 40 minutes. Serve packages unwrapped and let everyone open his own present. Serve hot as a luncheon dish.

Serves 8 to 12

*Available in Oriental markets.

**Available in some Chinese markets. Substitute corn husks or cabbage leaves.

Cantonese Fried Chicken

The rice flour makes an excellent, crisp, light coating for this all-time favorite.

3 to 4 pounds fryer chicken parts
2 tablespoons soy sauce
2 garlic cloves, finely minced
½ teaspoon finely minced ginger root
1 teaspoon salt
½ teaspoon white pepper
1 egg, beaten
¾ cup rice flour
Vegetable oil for deep-frying

Marinate chicken in soy, garlic and ginger 2 hours. Season with salt and pepper and combine with egg. Gradually sprinkle in the flour, coating chicken pieces with the thick "batter". Heat oil in wok or heavy skillet. Fry a few pieces of chicken at a time, being careful not to overcrowd, 8 minutes on each side or until golden. Drain on rack. Keep warm in oven until all pieces have been fried.

Serves 4 to 6

Variations:
- Herb Fried Chicken. Omit the soy, garlic (optional) and ginger. Add 2 tablespoons minced chives and 1 teaspoon minced tarragon leaves to chicken. No need to marinate.
- Rosemary Fried Chicken. Add a generous fresh sprig of rosemary to oil while heating. Remove before frying chicken.

Fragrant Beef With Rice Crumbs

A very popular Szechuan method of preparing steamed meat dishes using rice crumb coating.

Rice crumbs: Place 1 cup long grain brown or white rice in a dry skillet with 1 whole star anise and 1 teaspoon crushed peppercorns. Stir over low heat until rice grains are lightly toasted 5 to 7 minutes. Remove from heat and let cool. Put into blender and blend to texture of bread crumbs. Store in covered jar.

1 pound flank or skirt steak
2 tablespoons minced green onions
1 teaspoon minced ginger root
½ teaspoon Five-spice powder
2 tablespoons soy sauce
2 tablespoons rice wine or sherry
1 tablespoon peanut oil
1 teaspoon Oriental sesame oil
1 teaspoon sugar
1 teaspoon salt
½ teaspoon pepper
Pinch cayenne pepper
1/3 cup rice crumbs (see above)

Cut meat into thin slices ⅛-inch thick and 1½ inches square. Combine remaining ingredients except rice crumbs and marinate meat 1 hour. Dredge each slice of meat in rice crumbs and place in shallow heatproof bowl. Place the bowl in a large pot or steamer over 2 inches of water, cover and steam over high heat 45 minutes or until rice crumbs are tender, being sure to retain enough water in pot to create plenty of steam and moisture to cook the rice crumbs thoroughly.

Serves 4

Note: Lamb or pork may be substituted for the beef. Pork spareribs, cut up, are especially delicious prepared in this way; increase steaming to 15 minutes.

Hom Joong (Chinese Tamales)

Traditionally, Hom Joong are eaten during the Chinese Dragon Boat Festival on the 5th day of May. These tasty rice Ti-wrapped packages are a meal in one. Accompany with a good Chinese tea.

6 cups glutinous rice (sweet rice)
½ cup shelled raw peanuts
1 cup mung beans or dried green peas
2 teaspoons salt
½ pound Lop Chiang (Chinese sausage),* cut in 1½-inch pieces
½ pound raw ham, cut in 12 slices
¼ pound salt pork, blanched and diced
6 salted duck eggs*, yolks only, halved
1 dozen dried chestnuts, soaked overnight
8 forest mushrooms, soaked to soften and thickly sliced
½ cup dried shrimp*, soaked to soften and drained
½ pound Ti leaves* or corn husks, washed, soaked overnight, rinsed and dried

Wash rice in several waters until liquid runs clear. Let soak in cold water overnight. Drain and combine with peanuts, mung beans and salt. Have remaining ingredients ready. Take two Ti leaves, sides overlapping to form a wider surface to hold filling. Place leaves in palm of your hand and fill with ¼ cup of rice mixture. Add a piece of each of remaining ingredients and then another ¼ cup of rice mixture. Place another Ti leaf over rice. Bring ends up, folding sides over to securely enclose filling. Bring ends down and fold over. Tie with string around entire package twice, allowing a little room for the expansion of rice. Repeat with rest of ingredients. Drop into a large pot of boiling water to completely cover the joong. Cover and cook over medium heat 6 hours. If made ahead, reheat in hot water 20 minutes. Serve hot.

Makes 1 dozen

*Available in Chinese markets

Steamed Crabs With Glutinous Rice

1½ cups glutinous rice (sweet rice)
2½ cups water
2 large live crabs, blanched in boiling water 1 minute, cleaned and cracked
2 teaspoons minced ginger root
2 green onions, chopped
2 tablespoons rice wine or sherry
1 tablespoon soy sauce
1 tablespoon oyster sauce*
1 tablespoon peanut oil
½ teaspoon Oriental sesame oil
½ teaspoon sugar
Salt and white pepper to taste
Coriander springs for garnishing

Wash and soak rice in water to cover 30 minutes. Drain, add 2½ cups water, bring to boil, lower heat to medium and cook until all bubbles disappear from surface of rice. Cover with tight-fitting lid, lower heat to simmer and cook 20 minutes. While rice is cooking, marinate prepared crabs in remaining ingredients except coriander. Transfer cooked rice to heatproof bowl. Top with crabs and pour marinade over. Steam above briskly boiling water 30 minutes. Garnish with coriander and serve immediately.

Serves 6

*Available in Oriental markets

Chirashi Sushi

This vinegared rice can be garnished with a variety of vegetables, seafood, egg garni, etc. Chirashi sushi is never molded or rolled. It is served cool or warm as a main dish.

1 recipe Sushi Rice (page 31)
1 cup dashi stock*
1 cup julienne-cut carrots
1 cup julienne-cut bamboo shoots
½ cup fresh peas or julienne-cut green beans
4 dried forest mushrooms, soaked to soften and thinly sliced
1 tablespoon Japanese soy sauce
1 tablespoon sake (rice wine)

2 eggs, beaten with
1 tablespoon dashi stock
1 teaspoon vegetable oil
1 or 2 sheets laver (nori)**
½ pound cooked shrimp
½ pound cooked crab meat
2 tablespoons slivered pickled red ginger**

Have ready the sushi rice dressed with vinegar. Heat the dashi, add carrots and cook 5 minutes. Remove with slotted spoon and set aside. Add bamboo shoots and cook 2 minutes. Remove and set aside. Add peas or beans and cook 3 minutes. Remove and set aside. Add mushrooms, soy and sake and cook 10 minutes. Let cool in stock and drain. Reserve all but 1 tablespoon of stock for another use. Make egg garni by heating an omelet

pan and brushing with oil. Add just enough of the eggs beaten with stock to cover surface of pan. Let set, carefully turn over and cook 1 minute. Remove from pan and repeat until all egg is used. Let cool and cut into thin strands about 2 inches long. Hold laver over a candle or low flame to toast and crisp lightly. Set aside. Place rice on a large lacquer tray or platter. Artistically mound each of the separate vegetables and seafood in alternate color patterns on top. Sprinkle egg garni, ginger and crumbled laver over all.

Serves 6

Dashi stock is made from kombu, a seaweed, and katsuobushi, dried bonito fish. Use the instant variety available in Japanese markets, Dashi No Moto.

**Available in Japanese markets.*

Sekihan (Red Rice)

In Japan, sekihan is often made for festive occasions such as weddings, birthdays or New Year's. The color red represents joy, light and vitality. Rice represents prosperity.

1 pound mochi gome (glutinous rice)
1 cup azuki*
Salt
2 tablespoons toasted black sesame seeds

Wash the rice until water runs clear. Soak in water to cover 4 hours. Drain and set aside. Wash beans in colander and put into a pot with 4 cups water. Bring to rapid boil, lower heat and simmer, uncovered, 45 minutes or until beans are tender but intact, adding more water if needed. Drain and reserve liquid. Combine the rice and beans in a pot. Measure bean water and add additional water to cover surface of beans and rice by 1 inch. Bring to boil over high heat. Reduce heat to medium and cook until all water has evaporated. Lower heat to simmer, cover and cook 20 minutes. Season with salt to taste and serve with sprinkling of black sesame seeds.

Serves 6

Red beans available in Oriental food stores and some health food stores.

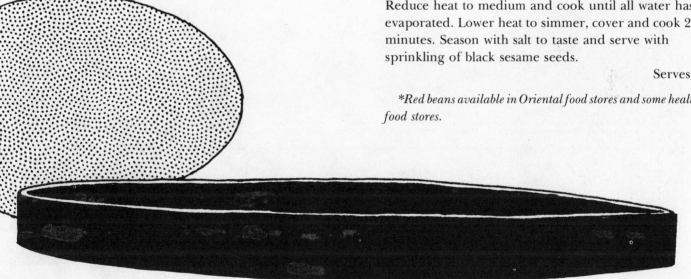

Domburi

Domburi are Japanese "one-bowl" rice dishes — plain steamed rice is topped with various ingredients. The following Domburi variations require freshly cooked Japanese rice on hand and Domburi Dipping Sauce.

DOMBURI DIPPING SAUCE

1 cup dashi stock*
⅓ cup Mirin (sweet sake) or ¼ cup dry sherry
⅓ cup Japanese soy sauce

Combine ingredients and bring to rapid boil. Let cool.

Makes approximately 1½ cups

For a slightly sweeter sauce add 1 tablespoon sugar to above ingredients.

OYAKO DOMBURI (CHICKEN, MUSHROOMS AND EGG)

1 sheet laver (nori)**
1 chicken breast, skinned, boned and thinly sliced
1 cup sliced fresh mushrooms
4 green onions, halved lengthwise and cut into
   2-inch lengths
Domburi dipping sauce
4 eggs
Hot cooked rice

Hold laver over candle or low flame to toast and crisp lightly. Cool, crumble and set aside. Divide chicken, mushrooms and onions into 4 portions and pour ¼ cup of the Domburi dipping sauce over each portion. In small omelet pan, heat one portion at a time until chicken is cooked. Beat 1 egg slightly, add to pan and cook until set. Immediately place over an individual bowl of hot rice. Sprinkle with reserved laver and serve immediately. Repeat with other portions.

Serves 4

TENDON DOMBURI (TEMPURA)

Allow 3 or more large tempura prawns per bowl of rice and any number of tempura vegetables of choice. Pour ¼ cup Domburi sauce over each portion and sprinkle with crumbled laver as above.

*Dashi stock is made from kombu, a seaweed, and katsuobushi, dried bonito fish. Use the instant variety available in Japanese markets, Dashi No Moto.
**Available in Japanese markets.

Tempura

Tempura is the Japanese version of French fried food — light and fluffy in texture and if prepared correctly not greasy. Fresh fish filets, oysters, lobster meat and shelled shrimp or prawns may be used, as well as vegetables such as sliced carrots, zucchini, sweet potatoes, asparagus, mushrooms, eggplant, onions, green peppers and beans, and herbs such as watercress or parsley springs. The secret is to pat the chilled seafood and vegetables dry before dipping in freshly-made batter. The peanut or corn oil must be clean and at least 2 inches deep in the frying pot. It must be kept at a temperature of 350°. Test by dropping a bit of batter into the hot oil; if it turns light in 1 minute the oil is hot enough. Fry only a few ingredients at a time in order to keep the oil temperature at 350°. As soon as ingredients are delicately golden, remove immediately to rack or paper toweling and serve at once with Domburi dipping sauce.

BATTER

¾ cup flour
2 tablespoons rice flour or cornstarch
⅔ cup cold water
1 egg, beaten

Mix flour, rice flour or cornstarch, water and egg lightly. Do not try to beat out lumps smoothly. Dip prepared seafood and vegetables in batter as directed. Make only one batch of batter at a time.

One-Pot Chinese Dishes

The Chinese are fond of "all-in-one rice pot" dishes and often prepare these simple, tasty quick meals for lunch. The preparation is a good convenience food for single dwellers as well as families. Be sure to cook a minimum of 1 cup raw rice for these one-pot meals. The meat is placed on top of the rice (Chinese method page 20) as soon as the water and bubbles disappear. Cover with tight-fitting lid and simmer 15 to 20 minutes, letting the rice take on the marvelous flavors of the meat and seasonings.

BEEF AND RICE: Plan about ¼ pound lean ground beef per person. Combine meat, ½ teaspoon soy sauce, pinch of minced ginger root (optional), 1 teaspoon chopped green onions and salt and pepper to taste. Crumble meat on rop of rice.

LOP CHIANG AND RICE: These Chinese smoked pork sausages are excellent steamed on top of rice. Use no more than 2 sausages per cup raw rice or the fat in the sausages will make the rice too greasy. Slice sausages after cooking and eat with the rice sprinkled with soy sauce. The flavored rice crust is superb for nibbling.

HAM AND RICE: Slice or dice raw ham and place on top of simmering rice. ½ cup diced ham per cup of raw rice is a good proportion.

HAM, EGGS, PEAS AND RICE: Add ½ cup each ham and peas when bubbles disappear on rice surface. Just before ready to serve, beat 1 egg and stir into rice, cover for 2 minutes and serve with soy or oyster sauce.

CHICKEN, MUSHROOM, BAMBOO SHOOTS AND RICE: Slice and combine 1 cup raw chicken meat, 2 forest mushrooms, soaked to soften, ¼ cup bamboo shoots, 1 tablespoon chopped green onions and 1 teaspoon each soy sauce, sherry and vegetable or peanut oil.

Fried Rice

Called "nasi goreng" by the Indonesians, fried rice is probably the rice dish best known to Westerners. The culinary artist may create a rice monument for the gods or merely a simple, earthy combination for his palate.

4 cups hot or cold cooked rice
¼ cup peanut oil
1 or more cups cooked meat or seafood, diced
1 cup fresh peas
2 tablespoons soy sauce
1 teaspoon sugar
Salt and pepper to taste
2 eggs, beaten
½ cup chopped green onions

Heat 2 tablespoons of the oil in wok or skillet and stir-fry meat and peas until peas are just cooked through. Do not overcook. Peas should be crisp. Remove from heat and turn out onto warm plate. Heat remaining oil in same skillet. When it sizzles, add rice, soy, sugar, salt and pepper, stirring constantly to blend and prevent rice from sticking. When thoroughly heated (if using hot rice, there is no waiting), add eggs. Cook and stir until eggs have set and are well mixed into the rice. Toss in the meat, peas and green onions.

Serves 4

"No Fry Fried Rice" variation: Stir-fry all ingredients except eggs and blend well into the pot of hot freshly cooked rice. Drizzle eggs into rice, cover and simmer 2 minutes. Stir well and serve.

HINT: Any of the following ingredients will glorify fried rice.

- Bamboo shoots, water chestnuts, peanuts, bean sprouts, carrots, celery, fresh or dried forest mushrooms, fresh coriander.
- Shrimp, scallops, abalone, lobster, crab, mussels or clams.
- Chicken, duck, turkey, pork, lamb, beef, ham, bacon, lop chiang (Chinese sausage).
- Oyster sauce, hoisin sauce, Oriental sesame oil, pinch of Five-spice powder.

Stir-Fry Variations

STIR-FRY VARIATIONS

A stir-fried meat and vegetable dish with a bowl of hot, steaming rice makes a wholesome, delicious complete meal in minutes. One could eat a variation on this theme for every day of the week and still be pleasantly surprised with every meal.

Stir-Fried Beef and Asparagus

1 pound flank steak, cut diagonally across the grain into thin strips
1½ pounds asparagus, sliced diagonally ¼ inch thick
1 onion, sliced
1 thin slice ginger root, minced
1 garlic clove, minced (optional)
1 tablespoon soy sauce
1 tablespoon sherry or rice wine
½ teaspoon sugar
3 tablespoons peanut oil
1 tablespoon cornstarch dissolved in
¼ cup cold water
Hot cooked rice

Combine beef, onion, ginger root, garlic, soy, sherry or rice wine and sugar. Let stand 10 minutes. Over high heat in wok or heavy skillet heat half the oil until it begins to smoke. Add the meat mixture and stir-fry constantly until meat begins to lose its redness. Immediately remove from heat and transfer to plate. Add remaining oil to wok. When hot, add the asparagus, sprinkle lightly with salt, stir to coat with oil, add ½ cup water and cover. Continue cooking 2 to 4 minutes until steam rises to top. Uncover, add meat mixture, stir to blend, and thicken with cornstarch mixture. Serve immediately over hot cooked rice.

Serves 4

Any of the following vegetables or combinations may be prepared in the same way:

- Broccoli, cauliflower, string or long beans, water chestnuts, mushrooms, bamboo shoots, carrots, zucchini, green pepper, pea pods, celery. Cut, slice or leave whole depending upon vegetable.
- Leafy vegetables such as spinach, cabbage or bok choy. Usually extra water is not needed.

Other meats and seafood may be substituted for the beef. Prepare boneless lamb as for beef. Pork and chicken should be cooked until done before removing from pan and proceeding with vegetable step. Shrimp, scallops, oysters, firm fish fillets, squid require quick, short cooking for best results.

Rice and Barley

In many Far East countries, when the rice supply is low, other grains play an important role in the daily diet of the natives. In Korea, rice and barley are often combined in the rice bowl.

1½ cups short or medium grain rice
½ cup pearl barley
4 cups water

Pour 2 cups boiling water over barley and soak overnight. Bring to boil, cover, lower heat and simmer 30 minutes. Wash rice until water runs clear, drain and combine with the barley and 2 cups water. Bring to boil, lower heat to simmer, cover with tight-fitting lid and cook another 20 minutes or until rice is tender.

Serves 4

Variation: Rice, Barley and Dates. To cooked rice mixture add ½ cup chopped pitted dates. Cover and let stand 10 minutes before serving.

Nasi Goreng (Indonesian Fried Rice)

6 cups cooked rice
3 tablespoons peanut oil
1 onion, chopped
1½ teaspoons curry powder
1 teaspoon finely minced ginger root
1 to 2 red chili peppers, minced
3/4 cup hot chicken stock
1½ tablespoons molasses
2 tablespoons soy sauce
½ cup chopped roasted peanuts
¼ cup chopped green onions and tops

Heat oil in large skillet or wok. Stir fry onion, curry powder, ginger root and chili peppers 3 minutes. Add rice, mix well and stir fry 5 minutes. Combine stock, molasses and soy sauce and blend into rice mixture. Continue cooking 2 minutes or until all liquid is absorbed. Remove from heat and toss in peanuts and green onions.

Serves 6

Khao Phat

From Thailand, this colorful fried rice is traditionally served in banana leaves.

4 to 5 cups cooked rice
3 tablespoons vegetable oil
1 tablespoon minced ginger root
1 garlic clove, minced
¼ pound lean pork steak, thinly sliced
½ pound raw shrimp, shelled and deveined
½ cup chopped green onions
2 eggs
Salt and pepper to taste
Banana leaves (if available)
Garnish of: **Fresh red or green chili peppers, thinly sliced cucumber and tomato slices**

In skillet or wok, heat oil and stir fry the ginger root and garlic 1 minute. Add pork and stir fry until well cooked. Add shrimp and cook and stir 2 minutes. Blend in rice and green onions and heat through. Into mixture crack 1 egg at a time, mixing well into the hot rice. Season with salt and pepper to taste. Mount rice mixture on banana leaves arranged on serving plate and garnish decoratively with chili peppers, cucumber and tomato slices. Serve warm.

Serves 4

Korean Style Rice With Vegetables And Beef

4 to 5 cups hot cooked rice
½ pound top round steak, cut in thin strips
3 tablespoons soy sauce
3 tablespoons peanut oil
¼ cup chopped green onions
2 garlic cloves, minced
2 tablespoons toasted, crushed sesame seeds
¼ teaspoon pepper
1 teaspoon sugar
1 cut thinly sliced carrots
½ pound bean sprouts
1 cup thinly sliced celery
1 cup thinly sliced cucumber (peeled if skin is
   tough)
½ cup thinly sliced water chestnuts
¼ cup water
Salt to taste

Garnish of:
1 green onion, slivered
egg garni
toasted, crushed sesame seeds

Combine meat, soy sauce, 1 tablespoon of the oil, green onions, garlic, sesame seeds, pepper and sugar. Let stand 15 minutes to blend flavors. Heat a heavy skillet or wok and stir fry the meat until redness disappears. Remove to plate. Add remaining oil to pan and stir fry carrots 2 minutes. Add bean sprouts, celery, cucumber and water chestnuts. Stir well and blend in water. Cover and let steam rise to surface, about 2 minutes. Vegetables should be heated through but remain tender crisp. Return meat mixture, mix well, season with salt and just reheat. Have rice ready in individual bowls or plates and top with vegetables and meat. Garnish and serve immediately.

Serves 4

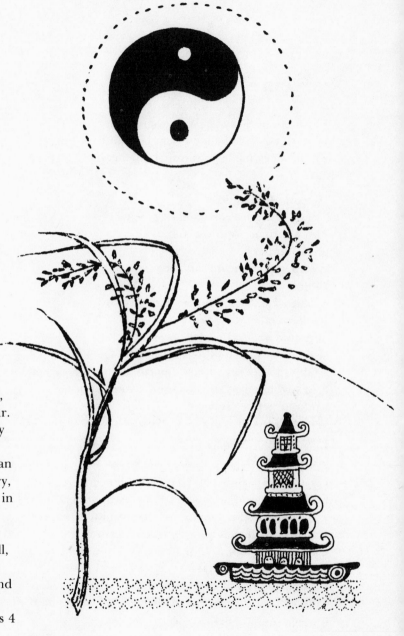

Arancini

Throughout Sicily, Arancini, large fried rice balls filled with meat and cheese, can be purchased in tavernas. They make an unusual luncheon fare served with red vino.

1½ cups Italian rice*
5 cups beef, veal or chicken stock
Pinch saffron (optional)
¼ cup butter
½ cup grated Parmesan cheese
2 eggs
3 tablespoons each butter and olive oil
1 onion, minced
½ pound ground veal or beef
¼ pound chicken giblets, minced
2 tablespoons each tomato paste and water, mixed
Pinch of crushed oregano and rosemary leaves
Salt and pepper to taste
½ cup diced Provolone or Mozzarella cheese
Fine bread crumbs
Oil for frying

Heat 3 cups of the stock to boiling point and gradually stir in rice. Continue cooking over medium heat until rice has absorbed all liquid. Add remaining stock and optional saffron which has been dissolved in a little stock. Cook 15 minutes, or until rice is tender and all liquid is absorbed. Add butter, Parmesan cheese and 1 of the eggs, beaten. Blend well and set aside to cool. For the filling, heat the butter and olive oil and saute onion until transparent. Add meats and, stirring constantly, cook until browned. Add tomato paste and water mixture, oregano and rosemary, and salt and pepper. Simmer 20 minutes, remove from heat and set aside to cool. To shape rice balls, form ¼ cup rice into a ball, make a hole in center and fill with 1 tablespoon of meat filling and 1 tablespoon diced cheese. Enclose hole and reshape ball. Repeat with remaining rice, meat filling and cheese. Dip balls into remaining egg, beaten, and roll in bread crumbs. Deep fry in hot oil until golden. Drain on paper toweling and serve hot. May be prepared ahead and warmed in a 350° oven. If serving as appetizers, form into smaller balls. Makes 8 to 10 large balls

Arborio or Vialone rice, available in Italian markets

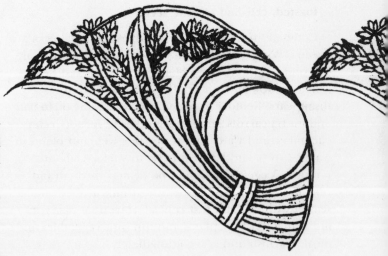

Boiled Wild Rice

The perfect accompaniment to wild game or fowl.

1 cup wild rice
3 cups water and/or stock
2 tablespoons butter
Salt to taste

Wash rice and soak in water to cover 30 minutes. Drain. Bring water and/or stock, butter and salt to boiling point in a 2-quart saucepan. Add rice without stirring, cover and cook over medium heat 30 minutes or until rice is tender. If rice has not absorbed all the water, drain and keep rice hot in a slow oven or over hot water.

Serves 4 to 6

Variations:

- Toss into hot cooked wild rice:
    1 cup chopped fresh mushrooms sauteed in 3 tablespoons butter, or
    ½ cup toasted, slivered blanched almonds, or
    ¼ cup crisply cooked bacon bits, or
    Pinch of herbs such as sage, rosemary, oregano, thyme, chives

- Last 10 minutes of cooking add 1 small onion, sauteed in 2 tablespoons butter
- Last 5 minutes of cooking add ¼ cup raisins or currants
- Combine cooked wild rice with cooked brown or long grain rice

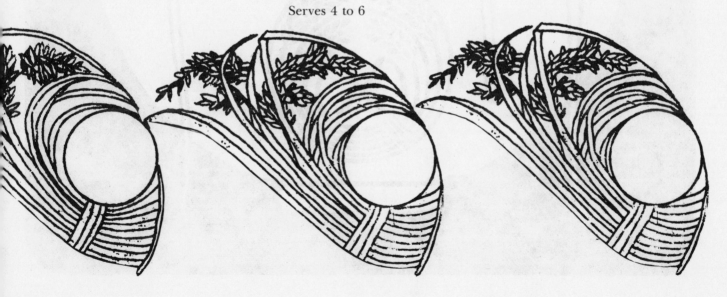

STUFFINGS & RICE NOODLES

Rice Stuffings

Rice stuffings are used in cuisines throughout the world where there are fowl to be dressed, "pockets" to be filled, or vegetables to be stuffed. Rice combines well with a great variety of herbs, spices and condiments, and the addition of fruits, nuts, smoked meats, vegetables and so on can make a rice stuffing a real enhancement to any meal.

Hints:

- Cooking the rice in stock will improve the flavor.
- Cool the stuffing before filling.
- Fill loosely to allow room for expansion during cooking.
- Any stuffings may be prepared as a rice dish to be heated and served separately.

Simple Rice Stuffing

2 cups cooked rice
¼ cup each chopped onion, celery and parsley
4 tablespoons butter
⅛ teaspoon each thyme, cloves and nutmeg
Salt and freshly ground pepper to taste

Sauté onions, celery and parsley in butter 5 minutes. Combine with remaining ingredients and let stand 1 hour to blend flavors.

Yield: 2½ cups

Variations:

- **Apple and Rice:** Sauté 2 cups peeled, chopped tart apples with the vegetables.
- **Apple, Sausage and Rice:** Omit butter and sauté ½ pound sausage meat in its own fat with vegetables until meat is thoroughly cooked. Drain off excess fat and combine with ingredients for Apple and Rice stuffing, adding ½ teaspoon sage.
- **Sweet Potato and Rice:** Add 2 cups mashed cooked sweet potatoes, ½ cup chopped pecans, 2 tablespoons brown sugar and 4 tablespoons crumbled cooked bacon (optional) to simple rice stuffing. Increase nutmeg to ½ teaspoon. Good for duck, chicken, turkey.
- **Apricot and Rice:** Add 1 cup chopped dried apricots and ¼ cup pine nuts when sautéing vegetables.
- **Ham, Mushroom and Rice:** Add 1 cup each chopped ham and fresh mushrooms when sautéing vegetables.
- **Cranberry and Rice:** Add 1 cup whole cranberry sauce and ½ cup chopped walnuts to simple rice stuffing. Good for thick pork chops or served with ham.

Rice and Oyster Stuffing

2 cups cooked rice
1 pint oysters
¼ cup chopped onion
4 tablespoons butter
¼ cup chopped celery
1 teaspoon grated lemon peel
⅛ teaspoon each thyme, nutmeg and cloves
½ teaspoon sage
Salt and freshly ground pepper to taste

Drain oysters, reserving liquor. Halve and sauté with onions in butter until onions are transparent. Add celery and oyster liquor and cook over medium heat until liquid is reduced by half. Remove from heat and cool. Combine with rice lemon peel and seasonings. Good stuffed in a whole fish or turkey.

Yield: 4 cups

Kidney and Rice Stuffing

2 cups cooked rice
4 lamb kidneys or 2 veal kidneys
1 garlic clove, finely minced
¼ cup each chopped celery, onion and parsley
4 tablespoons butter
⅛ teaspoon each rosemary, nutmeg and cloves
Salt and freshly ground pepper to taste

Sauté lamb or veal kidneys, garlic and vegetables in butter until kidneys are cooked but centers are still pink. Cool and slice kidneys ⅛ inch thick. Combine vegetables and kidneys with rice and seasonings. Stuff into boned leg of lamb or breast of veal.

Yield: 4 cups

Chinese Rice Stuffing

Hearty and delicious eaten as a stuffing or a luncheon meal.

6 cups cooked rice, kept hot
2 lop chiang*(Chinese sausages), finely minced
½ pound ground lean pork
¼ cup tiny dried shrimp*, soaked to soften
6 dried forest mushrooms, soaked to soften and minced
1 piece tangerine peel*, soaked to soften and minced
3 tablespoons peanut oil
½ teaspoon Oriental sesame oil
2 tablespoon each soy sauce, oyster sauce* and sherry
½ cup chicken stock or water
½ cup chopped green onions
2 tablespoons chopped fresh coriander
Salt to taste

Stirring constantly, stir-fry sausage, pork, shrimp, mushrooms and tangerine peel in peanut and sesame oil until pork looses its pinkness. Add soy, oyster sauce, sherry and stock or water. Cover and simmer 10 minutes. Combine with hot rice, making sure mixture is well blended, and toss in green onions and coriander. Season with salt. Serve immediately or cool and use as stuffing for turkey, chicken, etc.

Yield: 8 cups

Note: Chinese rice stuffing may be prepared using glutinous rice as well as short grain rice.

Available in Oriental markets.

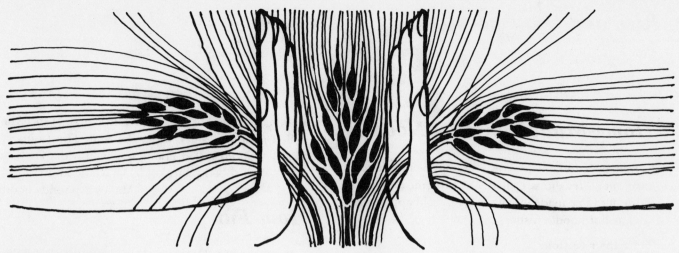

Mee Krob

From Thailand, fried rice vermicelli beautifully dressed with pork, shrimp, bean sprouts and colorful garnishes.

1 pound py mei fun*
Vegetable oil for deep frying
3 eggs, beaten
1 onion, thinly sliced
2 garlic cloves, minced
½ pound lean pork butt, cut in thin strips
1 pound raw shrimp, shelled and deveined
1 pound bean sprouts
2 tablespoons mien see**
1 tablespoon soy sauce
2 tablespoons lemon juice
Chopped green onions or chives
Chopped chili peppers
Coriander sprigs
Grated orange peel

In a wok or skillet, heat enough oil for deep frying. When hot add a few tablespoons of beaten egg at a time in a thin stream to make a lacy effect. Fry until just golden, remove with slotted spoon and repeat until eggs are used. Set aside. Without crowding fry a few handfuls of py mei fun at a time until just golden. They will expand greatly in volume. Keep warm in oven while frying remaining noodles. Empty all but 2 tablespoons of the oil from pan. Heat to sizzling and stir-fry the onion, garlic and pork until lightly browned. Add shrimp, bean sprouts, mien see, soy and lemon juice. Continue to stir-fry until meats are cooked, about 3 minutes. Combine with ¾ of the fried py mei fun and heap into a mound on large platter. Cover with remaining py mei fun. Top with the lacy fried egg and garnish with green onion or chives, chili peppers, coriander sprigs and grated orange peel. Serve hot.

Serves 6 to 8

*Rice vermicelli available in Oriental markets.
**Bean sauce. Also known as brown bean sauce or bean condiment. Available in Oriental markets.

Home-Made Fun (Rice Noodles)

In many Chinatowns, fresh fun may be purchased at take-out pastry shops. Called "bok fun", meaning white flour, it can be made into a variety of cold appetizers as well as hot noodle dishes.

2 cups rice flour
½ teaspoon salt
½ teaspoon powdered alum (to prevent sticking)
2¾ cups water
Vegetable oil

Place the rice flour on pastry board and sprinkle with salt and alum. Blend into flour. Gradually add a little water at a time with 3 tablespoons oil. Begin kneading, using only enough water to form a ball of dough. Knead 10 minutes to give an elastic, slightly chewy texture to the finished noodle skin. Place dough in mixing bowl, break up with hands and gradually add remaining water to make a batter. The mixture must be free of lumps. Let rest for 30 minutes. Have ready a large pot (wok works well) for steaming. Put 2 inches of water in pot or wok and place a rack or bamboo steamer over the water. Oil a pie pan or pizza pan, place on rack and ladle just enough batter in to cover surface of pan. The noodle will be as thick or thin as the amount of batter added. Cover, bring water to boil on high heat and reduce to medium as soon as it boils. Steam 15 minutes or until set. Wipe lid dry after each steaming. Have ready another oiled pan to repeat above process. After removing from heat, let cool just enough to handle and roll the noodle off the pan. Set aside on a plate until all noodles are made. For Chow Fun (stir-fry) slice the rolled noodles ½-inch wide. For Suey Fun (soup noodles) sliced rolled noodles ¼-inch wide. For Guon Fun (filled rolled pancake) leave whole and flat.

Makes 2 pounds noodles

Hom Fun

Prepare as for Home-made Fun, except after ladling each batch of batter onto pan, immediately sprinkle with 2 tablespoons of the following mixture. Proceed with the steaming as directed, cool, roll and slice in 1-inch lengths. Serve for lunch or as an appetizer.

1 cup finely minced ham or Chinese barbecued pork
½ cup tiny dried shrimp*, soaked to soften and squeezed dry (optional)
½ cup finely chopped green onions
3 tablespoons toasted sesame seeds
2 tablespoons finely minced choong toy* (preserved turnips*), optional
½ teaspoon Oriental sesame oil

Combine ingredients and mix thoroughly.

Available in Oriental markets.

Suey Fun (Soup Noodles)

Very popular with the Chinese for lunch or a midnight snack. Noodles represent longevity to the Chinese and are served for birthdays, especially honoring those who have reached a ripe old age. Never cut long noodles for these festive occasions.

1 recipe Home-made Fun (page 91) or
1 pound sha ha fun*
2 quarts chicken or beef stock
6 dried forest mushrooms, soaked to soften and
    halved if large
1 pound bok choy or spinach, washed and cut up
¾ to 1 pound Chinese barbecued pork, sliced
    coriander sprigs or chopped green onions

Heat stock and mushrooms, bring to boil and simmer
30 minutes. Add bok choy or spinach, bring back to boil
and add noodles and pork. When heated through, serve
immediately with garnish of coriander sprigs or
chopped onions.

Serves 6

Variations:

- Omit pork. With the vegetable add ¾ to 1 pound raw
  seafood such as shrimp, squid, abalone.
- Add sliced water chestnuts and/or bamboo shoots
  with the vegetable.

Chicken and Vegetable Chow Fun

1 recipe Home-made Fun or
1 pound sha ha fun*
1 pound chicken breast meat, sliced on diagonal
    ¼ inch thick
1 teaspoon sugar
1 teaspoon salt
½ teaspoon white pepper

1 teaspoon cornstarch
1 tablespoon each soy sauce and sherry
6 tablespoons peanut oil
4 forest mushrooms, soaked to soften and thinly
    sliced
½ pound fresh bean sprouts
½ cup sliced water chestnuts
1 onion, sliced
1 cup sliced celery
½ teaspoon Oriental sesame oil

Combine chicken with sugar, salt, pepper,
cornstarch, soy and sherry. Heat 2 tablespoons of the
peanut oil in wok or heavy skillet and stir-fry chicken 3
minutes, or until golden. Remove to plate. Heat another
tablespoon of oil in wok, add mushrooms and stir-fry 1
minute. Add vegetables and ¼ cup water. Cover and
cook until steam rises to top and vegetables are heated
through but still crisp. Transfer to plate. Heat
remaining oil in wok with sesame oil. Add fun or rice
sticks and using a wide spatula, lift noodles from bottom
to prevent sticking. Fun should be heated through and
slightly browned after cooking 5 to 7 minutes. If
noodles stick, add a little more oil to wok. Return
chicken and vegetables to wok, combining well with the
noodles.

Serves 6

*Individually coiled dried rice noodles sliced ¼ inch thick.
Soften in tepid water, undo coils and parboil 1 minute. Plunge
in cold water, drain and proceed as directed. Py mei fun (rice
vermicelli) may be used in the same manner.*

BREADS

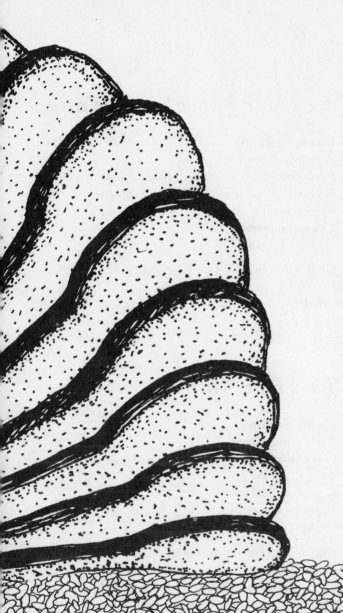

Rice Bread

A delicious, moist bread with a delicate scent of honey.

2 cups cooked, moist rice
2 tablespoons dry yeast
1½ cups lukewarm water
¼ cup honey
2 tablespoons corn or rice oil
2 teaspoons salt
½ cup rice bran
5 cups unbleached flour

In large bowl, mix yeast, water and honey until dissolved and frothy. Thoroughly blend in rice, oil, salt and rice bran. Add enough flour to form a soft dough. Knead dough on floured board at least 15 minutes until smooth and elastic. Form into ball, place in greased bowl, cover with damp tea towel and set in warm place to rise 1½ hours or until double in size. Punch down, let rise again 40 minutes or until doubled and return to floured board. Knead and shape into 2 loaves, place in 2 greased bread pans and let rise until again double in size. Bake in preheated 375° oven 10 minutes, lower heat to 325° and continue baking 35 minutes. Remove from oven, let stand 5 minutes, remove from pans and cool on rack.

Makes 2 loaves

Rice Pancakes

1 cup hot cooked rice
2 cups milk
3 tablespoons melted butter
1 teaspoon salt
4 eggs, separated
⅓ cup flour

Combine rice, milk, butter and salt. Beat egg yolks until light and blend into rice mixture. Blend in sifted flour, beat egg whites until stiff and carefully fold into rice mixture. Drop by tablespoons onto hot greased griddle. Cook, turning once, until golden on both sides. Serve with butter and maple syrup or honey, or with sweetened strawberries and sour cream, or with cinnamon-applesauce.

Makes 2 dozen 3-inch pancakes

Variation: Add ¾ cup chopped pecans or toasted almonds to batter.

Rice Flour Pancakes or Waffles

1½ cups rice flour
½ cup rice bran
1 teaspoon each sugar and baking soda
½ teaspoon each salt and baking powder
2 eggs, beaten
2 cups buttermilk
¼ cup vegetable oil

Sift dry ingredients. Combine eggs, milk and oil. Add to flour mixture, beating well until smooth. Bake on hot greased griddle or in waffle iron. Serve with butter and maple syrup.

Makes 2 dozen 4-inch pancakes or 8 waffles

Rice Flour Muffins

2 cups rice flour
2 tablespoons sugar
½ teaspoon each baking soda and salt
2 eggs, beaten
¼ cup melted butter
1¼ cups buttermilk
½ cup currants

Sift dry ingredients into bowl. Combine eggs, butter and buttermilk and stir into flour. Blend in currants. Fill greased muffin tins ⅔ full. Bake in preheated 400° oven 20 minutes.

Makes 1 dozen

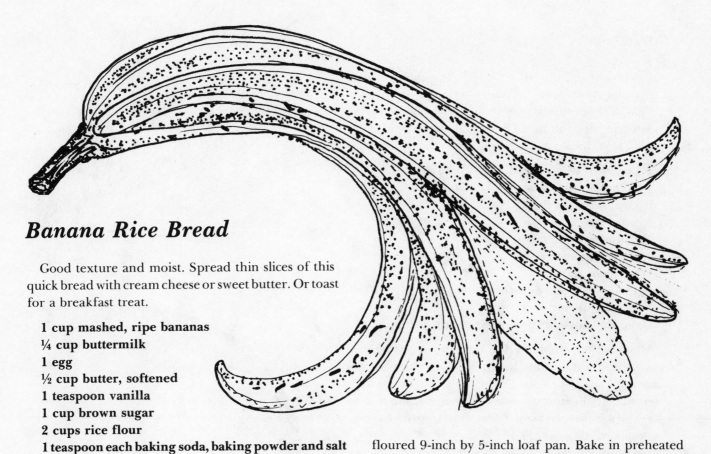

Banana Rice Bread

Good texture and moist. Spread thin slices of this quick bread with cream cheese or sweet butter. Or toast for a breakfast treat.

1 cup mashed, ripe bananas
¼ cup buttermilk
1 egg
½ cup butter, softened
1 teaspoon vanilla
1 cup brown sugar
2 cups rice flour
1 teaspoon each baking soda, baking powder and salt
1 cup chopped pecans

Place bananas, buttermilk, egg, butter, vanilla and brown sugar in blender and puree until smooth. Sift flour, soda, baking powder and salt and blend in a bowl with banana mixture, mixing with rubber spatula until smooth. Stir in pecans and pour into greased and floured 9-inch by 5-inch loaf pan. Bake in preheated oven 1 hour or until toothpick inserted in center comes out clean.

Carrot Rice Bread: Omit bananas and vanilla. Add 1 cup cooked, mashed carrots, 1 teaspoon cinnamon and ¼ teaspoon each cloves and freshly grated nutmeg.

Makes 2 loaves

Calas

At the turn of the century, the Creoles would fry calas from their carts on the streets of New Orleans to the delight of children and many passers-by. These can be called "rice doughnuts". They are often served for breakfast with café au lait.

1½ cups freshly cooked, moist hot rice
1 tablespoon dry yeast
½ cup warm water
3 eggs, beaten
1½ cups sifted flour
2 tablespoons sugar
½ teaspoon salt
¼ teaspoon freshly grated nutmeg
Vegetable oil for deep frying
Powdered sugar and cinnamon for dusting

Mash rice and cool to lukewarm. Dissolve yeast in warm water and blend well with rice. Cover and let rise in warm place overnight. Add eggs, flour, sugar, salt and nutmeg. Beat until smooth and let stand in warm place 30 minutes. Heat oil to 350° and drop in heaping tablespoons of batter. Fry 3 minutes or until golden, bouncing the calas in the oil as they are frying to increase their volume and make them lighter. Do not crowd. Drain on rack and dust with powdered sugar and cinnamon. Serve hot with or without maple syrup.

Makes 2 dozen

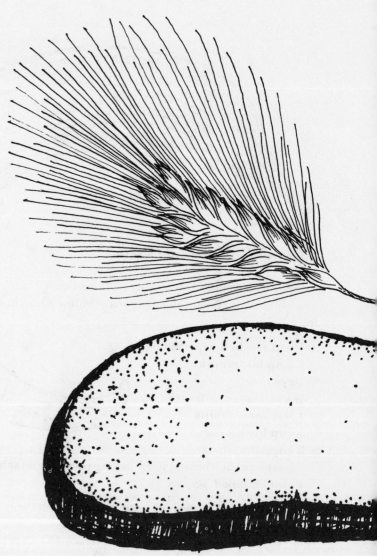

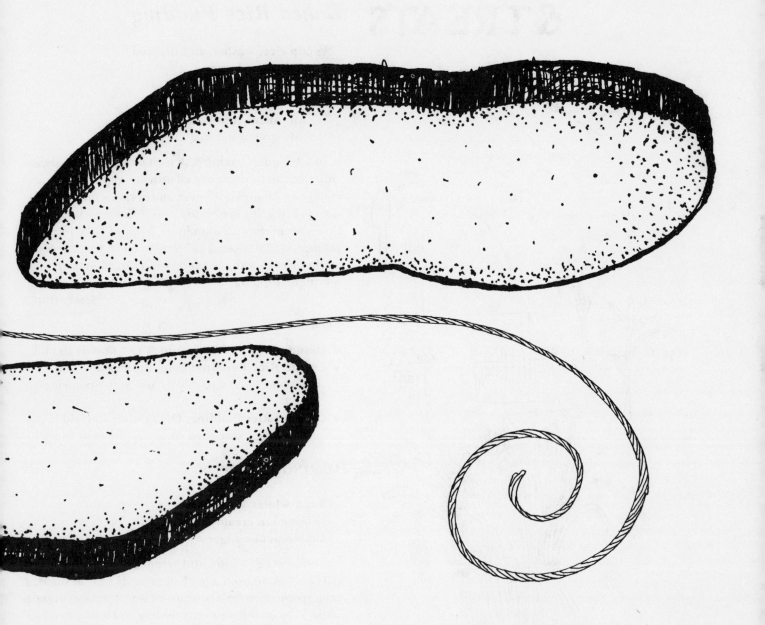

SWEETS &TREATS

Baked Rice Pudding

½ cup rice, washed and drained
3 tablespoons butter
4 cups milk
4 tablespoons sugar or honey
½ cup raisins (optional)
Freshly grated nutmeg

In a 1½-quart baking dish melt butter over low heat. Add rice and coat grains well with butter. Scald milk and combine with sugar or honey and optional raisins. Pour over rice, stirring constantly, and sprinkle with nutmeg. Cover securely with aluminum foil and bake in a preheated 250° oven 2½ hours or until firm. Serve warm or cold with cream or soft custard or with meringue topping.

Serves 6 to 8

Variations:

- **Brandy Rice Pudding:** Add ¼ cup brandy to milk.
- **Lemon Rice Pudding:** Omit nutmeg and add 1 teaspoon grated lemon peel and ¼ teaspoon lemon extract.
- **Coconut Rice Pudding:** Omit raisins and add ½ cup unsweetened shredded or grated coconut to milk.

MERINGUE TOPPING

3 egg whites
¼ teaspoon cream of tartar
6 tablespoons sugar

Combine egg whites and cream of tartar. Beat until stiff enough to hold a peak. Gradually beat in sugar, 1 tablespoon at a time. Continue beating until meringue is glossy. Pile onto hot baked pudding and return to a 425° oven 5 to 7 minutes or until topping is brown.

Chestnut and Rice Pudding

From Sicily, a tempting rice pudding with chestnuts and raisins.

1 cup dried chestnuts
1½ cups short grain rice
5 cups milk
½ teaspoon salt
½ cup sugar
½ cup white raisins
4 tablespoons butter
Freshly grated nutmeg (optional)

Soak chestnuts in water to cover overnight to soften. Drain, combine with milk and simmer 30 minutes. Add rice and salt and cook another 30 minutes. Add sugar and raisins, blend well and cook 5 minutes. Remove from heat, add butter and pour into bowl. Sprinkle with nutmeg and let cool before serving. Or grease bowl with butter, sprinkle with sugar and pour pudding into bowl. Let cool and chill until firm. Invert on platter and serve.

Serves 6

Variation: For a beautiful holiday pudding, add ½ cup mixed candied fruit and 2 tablespoons rum when adding butter.

Baked Almond Pudding

2½ cups cooked rice
3 eggs
¼ cup sugar
½ cup ground almonds
1 cup sour cream
1 tablespoon grated lemon peel

Beat the eggs until foamy, gradually add the sugar and continue beating until pale yellow and thick. Lightly fold in the ground almonds, sour cream, lemon peel and rice. Pour into buttered baking dish and bake in a preheated 325° oven 30 minutes or until pudding is lightly browned on top. Serve hot or cold with fruit sauce or lingonberry preserves.

Serves 4 to 6

Old Fashioned Rice Pudding

½ cup rice
¼ teaspoon salt
3 tablespoons sugar
¼ teaspoon cinnamon or nutmeg
2½ cups milk
½ cup raisins (optional)

Place all ingredients in top of double boiler. Cook, covered, over boiling water 1 hour or until rice is tender and milk is almost absorbed. Stir frequently to prevent rice from sticking to bottom of pot. Serve warm with cream or soft custard. Or dip a mold in water, shake out excess water and pour hot pudding into mold. Let cool and chill. Unmold onto serving plate.

Serves 6

Creme Caramel Au Riz

1½ cups cooked rice
1 cup sugar
2 tablespoons water
1 teaspoon lemon juice
2 eggs, beaten
2 cups milk
1 teaspoon vanilla

Combine sugar, water and lemon juice in saucepan and cook and stir over low heat until sugar is dissolved. Bring to boil and boil until syrup is golden brown. Pour into bottom and sides of an ovenproof dish and allow to set. Blend eggs, milk, vanilla and rice. Pour into caramelized dish and bake in a preheated 350° oven 45 minutes or until set. Let cool 10 minutes and invert onto plate so that caramel syrup is on top.

Serves 6

Swedish Rice Pudding

Traditionally served on Christmas Eve in Sweden, we are told that whoever finds the almond will be married before the next Christmas.

1 cup rice
3 tablespoons butter
1 quart milk
1 cinnamon stick
1 blanched almond
1 cup heavy cream
3 tablespoons sugar
1 teaspoon salt
1 teaspoon vanilla
1 cup raspberry preserves

Melt butter in saucepan and add rice and 1 cup water. Bring to boil and cook, uncovered, 5 minutes or until water has evaporated. Add milk and cinnamon stick and simmer, uncovered, 35 minutes, stirring occasionally. Remove from heat, gently stir in the almond, cream, sugar, salt and vanilla. Heat the raspberry preserves in top of double boiler. Serve the pudding hot with the raspberry preserves.

Serves 6 to 8

Riz A La Imperatrice

Aristocrat of rice desserts from France.

½ cup rice, washed and drained
2 cups milk
½ cup sugar
2 eggs, separated
1 teaspoon vanilla
1 tablespoon unflavored gelatin softened in
¼ cup water
¾ cup finely chopped mixed candied fruit
  marinated in
2 tablespoons kirsch
1 cup heavy cream, whipped
1 cup currant jelly
3 tablespoons slivered candied orange peel

Combine rice and milk and cook over medium heat, stirring frequently to prevent sticking, 35 minutes or until rice is tender. Stir in sugar and gradually blend ½ cup of mixture into egg yolks. Return to rest of rice mixture and continue stirring until thickened. Remove from heat and blend in vanilla, gelatin and candied fruit

and kirsch. Let cool. Whip egg whites until stiff and fold into rice mixture with whipped cream. Pour into a ring mold and chill until firm. Unmold onto chilled platter and surround with softened currant jelly mixed with kirsch to taste. Sprinkle with candied orange peel.

Serves 6

Note: If a smoother texture is preferred, after cooling mixture to lukewarm pour into blender and puree until smooth before proceeding with recipe.

Variation: Omit the candied fruit, currant jelly and kirsch. Marinate 1 quart fresh strawberries in ¼ cup Grand Marnier and fill center of mold.

Rice Torte

    1½ cups rice
    6 cups milk, or
    3 cups each evaporated milk and water
    1 cup sugar
    4 eggs, separated
    ½ cup blanched almonds, toasted and chopped
    ¼ cup each diced candied citron and candied
        orange peel
    2 teaspoons freshly grated lemon peel
    1 teaspoon vanilla
    ½ teaspoon almond extract
    Butter
    Fine dry bread crumbs
    ¼ cup Maraschino cherry syrup
    Powdered sugar

Bring milk to boil, stir in rice and sugar and simmer gently 25 minutes or until rice is tender and mixture thick. Remove from heat and cool. Beat egg yolks until light and fold into rice mixture with the almonds, candied citron and orange peel, lemon peel, vanilla and almond extract. Beat the eggs whites until stiff and fold into rice mixture. Butter a 12-inch spring-form pan generously with butter and coat lightly with bread crumbs. Pour batter in and bake in preheated 325° oven 1½ hours or until toothpick inserted in center comes out clean. Cake should be firm, light and golden on top. Let cool to lukewarm in pan, prick top and drizzle Maraschino syrup onto cake. Cool in pan, turn out and dust with powdered sugar.

Serves 10

Rice Flour Crêpes

½ cup cold water
⅔ cup cold milk
2 eggs
¼ teaspoon salt
2 cups rice flour
2 tablespoons melted butter
Apricot preserves
Powdered sugar

Put water, milk and eggs in blender. Add flour and butter and blend 1 or 2 minutes until smooth, scraping down any flour that sticks to sides of blender. Cover and refrigerate at least 2 hours or overnight. Heat a 6- or 7-inch crêpe pan or skillet, add a small dab of butter and heat until bubbly. Stirring batter often, pour in about 3 tablespoons, tilting pan to cover bottom. Cook until just golden, turn and cook other side. Stack on rack until all batter is used. Place about 1 tablespoon of the preserves on one corner, fold in edges and fold over once. Place in buttered shallow casserole to reheat. Do not crowd. Dot with butter and heat in 350° oven 10 minutes or until heated through. Just before serving sprinkle with powdered sugar.

Makes 18 crêpes

These crêpes may be filled with anything you desire and may be flambéed with liqueur or brandy sauce.

Brandy Fruit Cookies

2 cups rice flour
1 teaspoon baking soda
½ teaspoon cinnamon
¼ teaspoon each nutmeg, powdered
  ginger and powdered cloves
½ cup rice bran
1 cup shortening
1 egg
⅓ cup evaporated milk
¼ cup brandy
½ cup honey
¼ cup brown sugar
1 teaspoon grated orange peel
½ cup chopped candied fruit
½ cup chopped nuts
Powered sugar

Sift flour, soda and spices and mix with bran. Cream shortening with egg, milk, brandy, honey, sugar and lemon peel, blending well until smooth. Gradually add rice flour mixture and stir in candied fruit and nuts. Drop by teaspoons onto greased cookie sheet and bake in preheated 350° oven 10 minutes. When cool dust with powdered sugar.

Makes 4 dozen

Date Nut Rice Cake

1 cup hot coffee
1 cup chopped pitted dates
½ cup butter or shortening
1 cup brown sugar
1 egg
1 teaspoon vanilla
1½ cups rice flour
1 teaspoon baking soda
½ teaspoon baking powder
½ teaspoon salt
½ cup chopped walnuts

Pour hot coffee over dates and let stand 15 minutes. Cream butter or shortening with sugar and egg until light. Add vanilla and coffee-date mixture. Sift dry ingredients and blend into creamed mixture. Fold in nuts and pour into greased 8-inch square pan. Bake in preheated 350° oven 50 minutes or until toothpick inserted in center comes out clean. Serve with plain or whipped cream.

Serves

Applesauce Rice Cake: Omit coffee, dates, and vanilla. Add 1 cup thick applesauce, ½ cup raisins, 1 teaspoon cinnamon and ¼ teaspoon each allspice and nutmeg.

Almond Tea

The Chinese delight in sipping tea. This version is really a light pudding, served in delicate porcelain tea cups with lids as a refreshing and smooth drink. Serve as a light dessert after an Oriental meal to be sipped or eaten with a porcelain spoon.

½ cup long grain rice, washed and drained
1½ cups blanched almonds
1 quart water
¼ cup sugar
½ teaspoon almond extract

Place rice and almonds in blender with enough water to puree and thoroughly pulverize the rice. Pour mixture and remaining water into a saucepan and cook over medium heat, stirring constantly, 25 minutes or until thickened. Add sugar and almond extract and pour into individual cups.

Serves 6

Jien Doy

Deep-fried rice balls with a sweet bean filling. Eaten as a pastry, jien doy are often served as one of the many "dim sum" dishes. They are a favorite after-school treat for Chinese children. Enjoyed throughout the Chinese New Year, jien doy are placed before ancestral portraits in many homes during this festive time.

2 cups glutinous rice flour
½ cup brown sugar
¾ cup water
1 cup dow sah*
½ cup sesame seeds
Vegetable oil for deep frying

Combine sugar and water and bring to rapid boil. Remove from heat and stir in rice flour, blending well to form a stiff dough. When cool enough to handle, form dough into ropes 1 inch in diameter. Cut into 1-inch pieces and roll into balls with palm of hands, coating

hands with a little oil if needed to prevent sticking. Roll the dow sah paste into ropes ½-inch in diameter. Pinch off ½-inch pieces and form into balls. Flatten a dough ball with palms of hands and place a dow sah ball in center. Bring up edges of dough to encase the dow sah. Form into a ball again, roll in sesame seeds, pressing seeds in well, and repeat with remaining balls. Fry in hot oil at least 3 inches in depth (wok works well) for about 10 minutes, bouncing the balls in the oil as they rise to the surface to create a greater volume of air in the balls to make them lighter and larger. Drain well on rack. Jien doy are best served hot, but may be eaten at room temperature.

Makes 2 dozen

Variation: In place of dow sah, make a filling of ¾ cup minced dates, ½ cup grated coconut and 2 tablespoons chopped roasted peanuts.

Sweet black bean filling, available in cans in Oriental markets. Sweet red bean paste may be substituted.

Gaw Shung Go (Nine-Layered Steamed Pudding)

On the third day of the Third Moon, Chin Ming, the Festival of the Tombs, is celebrated. The Chinese visit the graves of the ancestors and present offerings and food to the departed. Families partake of Gaw Shung Go which is significant of the Nine Complimentaries related to sacred power and ever-flowing life. Most pagodas have nine tiers and the Altar of Heaven, built in Peking, was constructed with mathematical genius in multiples of nine.

1 cup brown sugar
1 pound rice flour (about 3½ cups)
3 tablespoons white sugar
¼ cup peanut oil
¼ cup toasted sesame seeds

Combine brown sugar and 2 cups water, bring to boil and simmer 5 minutes. Cool and strain into mixing bowl. Stirring constantly until well blended, gradually add rice flour. Slowly stir in 1½ cups more water to the consistency of pancake batter. Beat in white sugar and oil. Let stand 20 minutes. Grease a deep 9-inch pie pan or 9-inch square pan generously with vegetable oil and place in large steamer with at least 3 inches of water (bamboo steamer and wok may be used). When water comes to boil and pan is hot, stir up batter and ladle ½ cup into pan, tilting to cover entire surface. Cover steamer and steam 5 minutes. The layer should be firm and cooked. Stir up batter again and ladle another ½ cup onto first layer. Wipe off any steam which may have accumulated on the lid. Cover and steam 5 minutes. Repeat this procedure until the last layer. After pouring in last batter place a cloth napkin over the steamer to prevent any moisture from falling on pudding. Cover and steam 15 minutes longer. Remove from heat and immediately sprinkle sesame seeds over surface. Cool to lukewarm before turning out onto board to cool. Cut into 1½-inch thick slices and serve at room temperature.

Serves 9 or more

Ohagimochi (Sweet Rice Cakes)

Small round rice cakes with a coating of red bean paste often served as dessert for elaborate Japanese parties.

2 cups mochi gome (glutinous rice)
1 cup sugar
2 tablespoons poppy seeds
4 cups red bean paste

Wash rice until water runs clear. Soak in water to cover overnight, drain and place in pot with 2 cups water. Bring to boil, reduce heat to medium and cook until water has evaporated. Lower heat to simmer, cover and continue cooking 20 minutes. Remove from heat. Add sugar and blend well. Cover and let stand until rice is lukewarm. Shape into walnut-size balls and cool. Put a heaping tablespoon of red bean paste on a damp, clean cloth. Press a rice ball in center of bean paste, bring up the cloth and press bean paste against rice ball to completely encase with a layer of bean paste. Twist cloth at peak. Unwrap, place on plate, smooth ball if necessary and top with poppy seeds. Repeat with remaining balls.

Serves 8 or more

RED BEAN PASTE

2 cups azuki (red beans)*
2 cups sugar
1 teaspoon salt

Wash beans and cover with 2 quarts water. Bring to boil and simmer until beans are very soft. Drain and rub through sieve. Strain through cheesecloth and squeeze out any excess moisture. Combine with sugar and salt. Cool before combining with rice balls.

Available in Japanese markets and some health food stores.

Sweet Congee

Eaten as a snack for any time of the day. Often served after a Mah-jong game.

½ cup glutinous rice (sweet rice)
1½ dozen lotus seeds;* soaked 30 minutes in hot water, drain
½ cup dried longans* or raisins
10 jujube dates*
2 tablespoons brown sugar or honey
Salt lightly to taste

Combine the rice, lotus seeds, longans, and dates with 3 quarts water. Bring to a boil and simmer 1½ hours or until thick and smooth. Stir occasionally to prevent any sticking on bottom of pot. If necessary, add additional boiling water during cooking. Add sugar or honey. Serve hot.

Serves 6

Available in Chinese food stores.

Puffed Rice

Serve lightly salted as a snack, or as a sweet treat sprinkled with sugar. Also good sprinkled over salads or as a garnish for Oriental dishes.

1 cup brown or long grain white rice
Vegetable oil for deep frying

Cook rice by Chinese method (page 20). Let the rice kernels dry thoroughly for a day or so until hard. Heat oil in small pot to 375° and put 3 tablespoons of rice at a time into hot oil. Cook until puffed. Drain on paper toweling and repeat with remaining rice.

Rice Tea

Wash and dry 1 cup rice. In a heavy skillet over low heat toast the rice grains until golden, stirring occasionally. Remove from heat, let cool and store in airtight container on cupboard shelf. To make a pot of tea: Add 1 tablespoon of the prepared rice to a warmed tea pot. Pour in 4 cups boiling water. Cover and let steep 10 minutes. Refreshing!

Genmai Cha (Japanese Rice Tea)

Prepare 1 cup of rice as for Rice Tea (preceeding). Combine the hot toasted rice with 1 cup Japanese green tea leaves. Cool, place in airtight container and store on cupboard shelf. To make a pot of Genmai: Add 1 tablespoon of the rice-tea mixture to a warmed tea pot. Pour in 4 cups boiling water. Cover and let steep 5 minutes.*

Finnish Rice Wine

3 pounds rice (not precooked variety)
1 orange with skin, thinly sliced
1 pound raisins
6 pounds white sugar
1 pound brown sugar
8 quarts boiling water
2 teaspoons dry yeast

Combine rice, orange slices, raisins and sugars in a large earthenware crock and pour boiling water over. Dissolve yeast in ½ cup warm water and blend with rice mixture. Cover with an earthenware lid and leave to ferment at room temperature 2 to 3 weeks or until wine has finished bubbling. stirring every morning and evening during the fermentation period. When ready, strain into sterilized bottles. If there is sediment, strain into other sterilized bottles until wine is clear. Cork and store in cool place.

*Genmai cha is made with roasted unhulled rice in Japan.

Kheer

A fragrant rice pudding from India.

1 cup rice
4 cups milk
2 cardamon pods, peeled and crushed
¼ cup sugar
2 teaspoons rose water
¼ cup pistachio nuts
Edible silver leaf (optional)

Put rice in blender to make fine crumbs. Bring milk just to boil, add rice and cardamon and cook over low heat, stirring constantly, 20 minutes or until thick. Add sugar and cook 1 minute. Remove from heat and blend in rose water. Pour into individual bowls and sprinkle with pistachios. Just before serving top with silver leaf. Serve hot or cold.

Serves 4 to 6

Puffed Rice Squares

7 cups puffed rice*
3 tablespoons toasted sesame seeds
½ cup butter
2 cups sugar
1 teaspoon vanilla

Spread puffed rice and sesame seeds in a shallow pan and heat in a very low oven while making syrup. Melt butter in saucepan over medium heat. Add sugar and vanilla, stirring constantly until sugar is dissolved and syrup begins to thicken, about 5 minutes. Pour over warmed rice-sesame seed mixture, blending quickly, and press mixture into a buttered 8-inch by 12-inch by 2-inch pan. Let cool and cut into 2-inch squares.

*Available in health food stores

Belgian Rice Tart

Serve for a special afternoon tea or coffee.

Butter pastry:
1½ cups sifted flour
¼ cup sugar
½ teaspoon grated lemon rind
Pinch of salt
5 tablespoons butter
1 egg, beaten with
1 tablespoon water

Combine flour, sugar, lemon rind and salt in mixing bowl. Cut in the butter with a pastry blender. Gradually blend in eggs with a fork. Form into ball, wrap in wax paper and chill. Roll out on lightly floured board and line a 9-inch tart or pie pan. Set aside.

Filling:

1/3 cup rice
2 cups rich milk
½ cup sugar
½ teaspoon salt
3 tablespoons butter
2 eggs, separated
½ teaspoon vanilla or almond extract
1 teaspoon grated lemon rind

Combine rice, milk, sugar and salt in top of double boiler. Cook, covered, over boiling water 1 hour or until rice is tender and mixture is thick. Stir often. Blend in butter and cool to lukewarm. Beat egg yolks with vanilla or almond extract and lemon rind. Combine with rice mixture and fold in stiffly beaten egg whites. Pour into prepared shell and bake in preheated 400° oven 35 to 40 minutes or until lightly golden and set.

Eight Treasure Rice Pudding

A glutinous Chinese rice and fruit molded pudding. Serve small quantities of this rich, elegant dessert with a choice tea as a beautiful dinner finale.

2 cups glutinous rice (sweet rice)
½ cup minced suet
¼ cup sugar
2 tablespoons lard
½ cup preserved kumquats, seeded and halved
½ cup honeyed dates, quartered lengthwise
¼ cup preserved or candied ginger, cut in thin strips
¼ cup candied cherries, halved
¼ cup candied orange peel, cut in thin strips
¼ cup candied pineapple, cut in ½-inch pieces

Wash rice and soak in water to cover 2 hours. Drain and place in saucepan with 3 cups water. Bring to boil, lower heat to simmer, mix in suet and sugar, cover and cook 20 minutes. Grease a 2-quart mold with the lard and cover with a thin layer of cooked rice. Press a portion of the fruits and nuts onto the rice in an attractive design, cover with layer of rice and repeat layers, ending with rice layer. Cover mold with aluminum foil to catch excess steam, place in large steamer above water, cover pot and steam over briskly boiling water 45 minutes. Invert pudding onto warm plate and serve steaming hot.

Serves 8 to 10

INDEX